# How the Heart Develops

*A Visual Approach*

# Colloquium Series on the Cell Biology of Medicine

Editor
**Joel D. Pardee**, Weill Cornell Medical College

**Published**

Skeletal Muscle & Muscular Dystrophy: A Visual Approach
Donald A. Fischman
2009

How the Heart Develops: A Visual Approach
Donald A. Fischman
2009

**Forthcoming Books**

The Body Plan: How Structure Creates Function
Joel D. Pardee
2009

Bones: Growth, Strength, and Osteoporosis
Michelle Fuortes
2009

Breast Cancer and the Estrogen Receptor
Joel D. Pardee
2009

(Forthcoming Books continued on page 61)

How the Heart Develops: A Visual Approach
Donald A. Fischman
www.morganclaypool.com

ISBN: 9781615040001 paperback

ISBN: 9781615040018 ebook

DOI: 10.4199/C00001ED1V01Y200904CBM001

A Publication in the Morgan & Claypool Life Sciences series

*COLLOQUIUM SERIES ON THE CELL BIOLOGY OF MEDICINE*

Book # 1

Series Editor: Joel D. Pardee, Weill Cornell Medical College

**Series ISSN** TBD

# How the Heart Develops
*A Visual Approach*

**Donald A. Fischman**
Weill Cornell Medical College

*COLLOQUIUM SERIES ON THE CELL BIOLOGY OF MEDICINE #1*

MORGAN & CLAYPOOL LIFE SCIENCES

## ABSTRACT

With possible exception of the atomic clock, the heart may be the most perfect machine ever devised. How it develops from a simple embryonic tube is a fascinating story of biology and lends a great deal of insight into the source of heart defects that affect children and adults alike.

Central to this entire lecture is the fact that the fetus resides in an aquatic environment. Oxygenated blood arrives from the placenta and deoxygenated returns to the placenta (Figure 1). The lungs play no role in delivering oxygen or removing carbon dioxide to or from the circulation. Thus, the fetus mainly (but not exclusively) requires a three-chambered heart rather than the four-chambered heart that we are all familiar with. This resembles fish circulation in which blood leaves the heart into an aortic sac from which emanate the aortic arches that deliver blood to the gills, where it is oxygenated and $CO_2$ is removed. Blood then goes to the dorsal aortae for nourishing the body tissues. In a fish there is no need for a four-chambered heart, since fish do not use lungs to aerate the blood or remove $CO_2$ (Movie 1). Although the fetus lacks gills and still develops a four-chambered heart, much of fetal circulatory physiology depends on a "quasi-three-chambered circulation" that bypasses the pulmonary circulation. Upon birth, this "aquatic" circulation must change within minutes to permit lung function. The topics to follow trace how this circulation develops and how it changes upon birth.


ATTENTION READERS OF THE PAPERBACK VERSION:
To view the video files associated with the digital version of How the Heart Develops, please use the following URL: http://www.morganclaypool.com/r/heart.


## KEYWORDS

overview of embryonic and fetal circulation, stages of embryonic and fetal hematopoiesis, formation of the tubular heart, origin of myocardium, endocardium, epicardium, conducting system, proepicardium, looping of the heart, pericardial cavity and transverse pericardial sinus, endocardial cushions and relationship to valve formation, septation of ventricles, septation of atria, partitioning of the outflow tract, cardiac neural crest

# Contents

Introduction to the Cell Biology of Medicine .......... ix

Introduction .......... 1

Fetal and Embryonic Hematopoiesis .......... 5

Formation of the Precardiac Mesoderm and Fate Mapping During Gastrulation .......... 9

Tubular Heart .......... 13

Cardiac Looping .......... 17

Pericardial Cavity .......... 22

Endocardial Cushions .......... 24

Atrial Septation .......... 26

Ventricular Septation .......... 32

Partitioning of the Bulbus Cordis and Truncus Arteriosus .......... 35

Conducting System .......... 38

Cell Lineages During Heart Development .......... 39

Circulation at Term and Changes upon Birth .......... 43

Recommended Readings .......... 53

Series Editor Biography .......... 55

Index .......... 57

## Forthcoming Books

Colloquium Series on the Cell Biology of Medicine ........ 61

Colloquium Series on the Integrated Systems Physiology........ 65

Colloquium Series on the Developmental Biology ........ 68

# Introduction to the Cell Biology of Medicine

In order to learn, we must be able to remember, and in the world of science and medicine, we remember what we envision, not what we hear. It is with this essential precept in mind that we offer the Cell Biology of Medicine lecture series. Each lecture is given by faculty accomplished in teaching the scientific basis of disease to both graduate and medical students. In this modern age, it has become clear that everyone is vastly interested in how our bodies work and what has gone wrong in a disease. It is likewise evident that the only way to understand medicine is to engrave in our mind's eye a clear vision of the biological processes that give us the gift of life. In these lectures, we are dedicated to holding up for the viewer an insight into the biology behind the body. Each lecture demonstrates cell, tissue, and organ function in health and disease, and it does so in a visually striking style. Left to its own devices, the mind will quite naturally remember the pictures. Enjoy the show.

Joel Pardee
New York, NY

# Introduction

In this lecture on the developing human heart, the following topics are covered:

1. Overview of embryonic and fetal circulation
2. Stages of embryonic and fetal hematopoiesis
3. Formation of the tubular heart
4. Origin of myocardium, endocardium, epicardium, conducting system
5. Proepicardium
6. Looping of the heart
7. Pericardial cavity and transverse pericardial sinus
8. Endocardial cushions and relationship to valve formation
9. Septation of ventricles
10. Septation of atria
11. Partitioning of the outflow tract
12. Cardiac neural crest

With possible exception of the atomic clock, the heart may be the most perfect machine ever devised. How it develops from a simple embryonic tube is a fascinating story of biology and lends a great deal of insight into the source of heart defects that affect children and adults alike.

Central to this entire lecture is the fact that the fetus resides in an aquatic environment. Oxygenated blood arrives from the placenta and deoxygenated returns to the placenta (Figure 1). The lungs play no role in delivering oxygen or removing carbon dioxide to or from the circulation. Thus, the fetus mainly (but not exclusively) requires a three-chambered heart rather than the four-chambered heart that we are all familiar with. This resembles fish circulation in which blood leaves the heart into an aortic sac from which emanate the aortic arches that deliver blood to the gills, where it is oxygenated and $CO_2$ is removed. Blood then goes to the dorsal aortae for nourishing the body tissues. In a fish there is no need for a four-chambered heart, since fish do not use lungs to aerate the blood or remove $CO_2$ (Movie 1). Although the fetus lacks gills and still develops a four-chambered heart, much of fetal circulatory physiology depends on a "quasi-three-chambered circulation" that bypasses the pulmonary circulation. Upon birth, this "aquatic" circulation must change within minutes to permit lung function. The topics to follow trace how this circulation develops and how it changes upon birth.

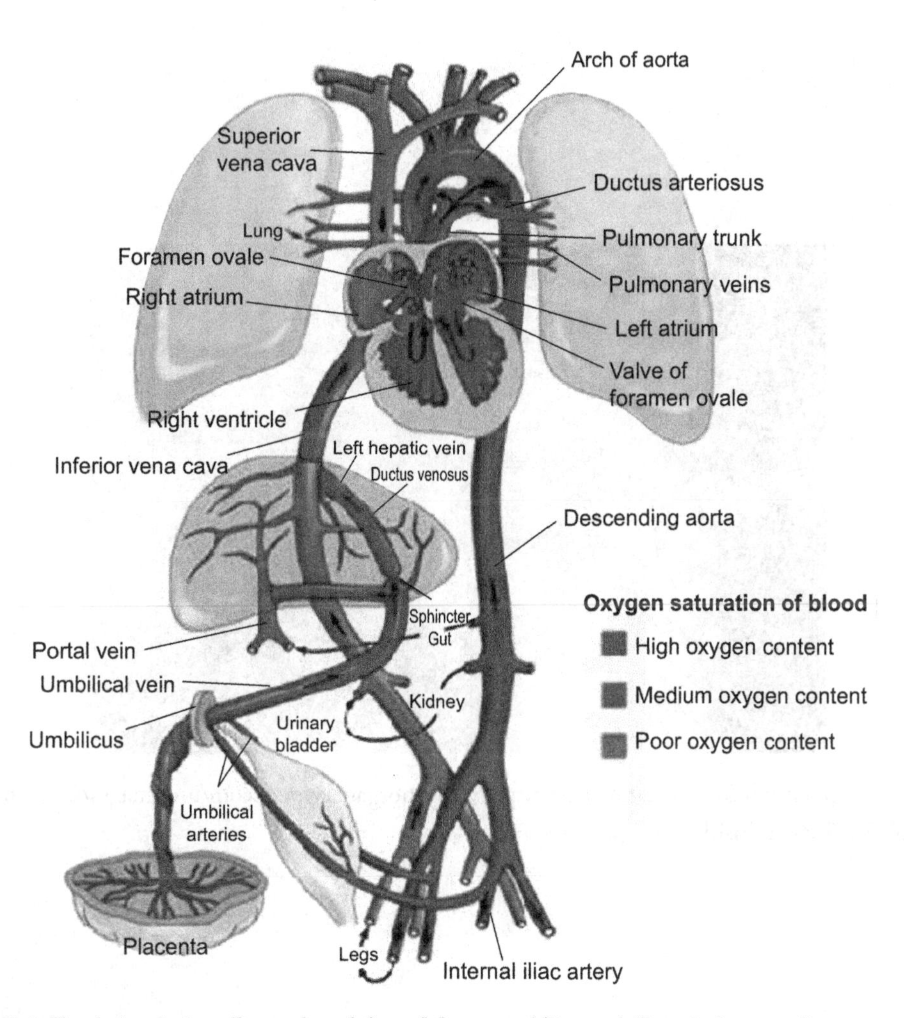


FIGURE 1: Fetal circulation. Reproduced from Moore and Persaud. Permission pending.

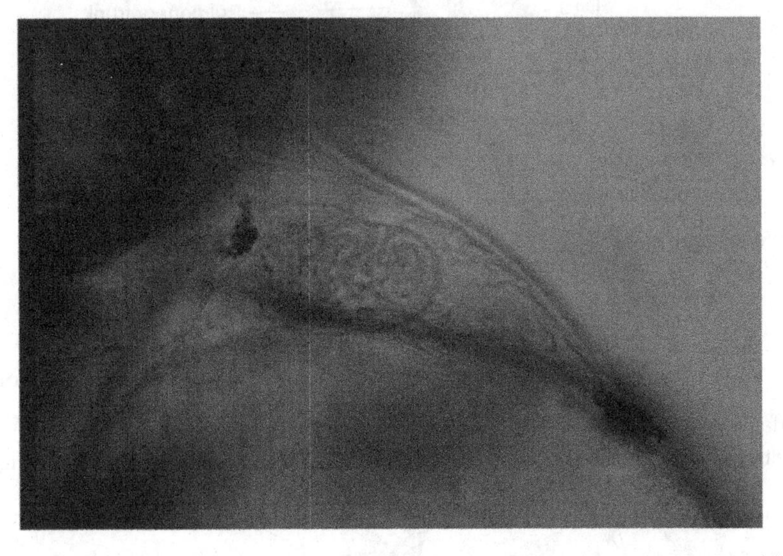

**MOVIE 1:** Zebra fish heart circulation (see http://www.morganclaypool.com/userimages/ContentEditor/1245101601659/ZebraFishHeartCirc.mov).

# Fetal and Embryonic Hematopoiesis

Initial blood cell formation begins in the splanchnic layer of the extraembryonic mesoderm surrounding the secondary yolk sac at approximately 20–21 days of embryonic development. (Figure 2). This mesoderm surrounds the epithelial cells of the yolk sac, and current evidence indicates that this tissue layer is derived from the yolk sac epithelium by a process of epithelial–mesenchymal transformation. Blood islands then form; these contain pluripotential hematopoietic stem cells. The blood cells enter the circulation on days 21–22 as the heart begins pumping. Blood vessels form concurrently by vasculogenesis (Figures 3–5). These yolk sac–derived erythrocytes are large nucleated cells, and for the first 6 weeks, the circulating red blood cells (RBCs) are almost entirely derived from the yolk sac. However, the cells from the yolk sac are soon replaced by cells from intraembryonic sites of hematopoiesis. The first sites of intraembryonic hematopoiesis are in small clusters of cells (paraaortic clusters) in the splanchnic mesoderm associated with the dorsal aorta and soon afterward appear in the aorta/genital ridge/mesonephros region (Figure 6). By 5 weeks, blood formation is evident in the liver. The RBCs from the liver are different in appearance and contain a different isoform of hemoglobin. Yolk sac synthesizes $\zeta$-Hb, also termed *embryonic Hb*. The liver-derived RBCs are nonnucleated and synthesize the fetal form of $\gamma$-Hb. By 6–8 weeks, the liver has replaced the yolk sac as the predominant site of hematopoiesis. Although the liver continues to produce blood cells into the neonatal period, its contribution declines after the sixth month of pregnancy. At this time, hematopoiesis shifts to the bone marrow; this shift appears to be regulated by cortisol, since in its absence, blood formation remains confined to the liver.

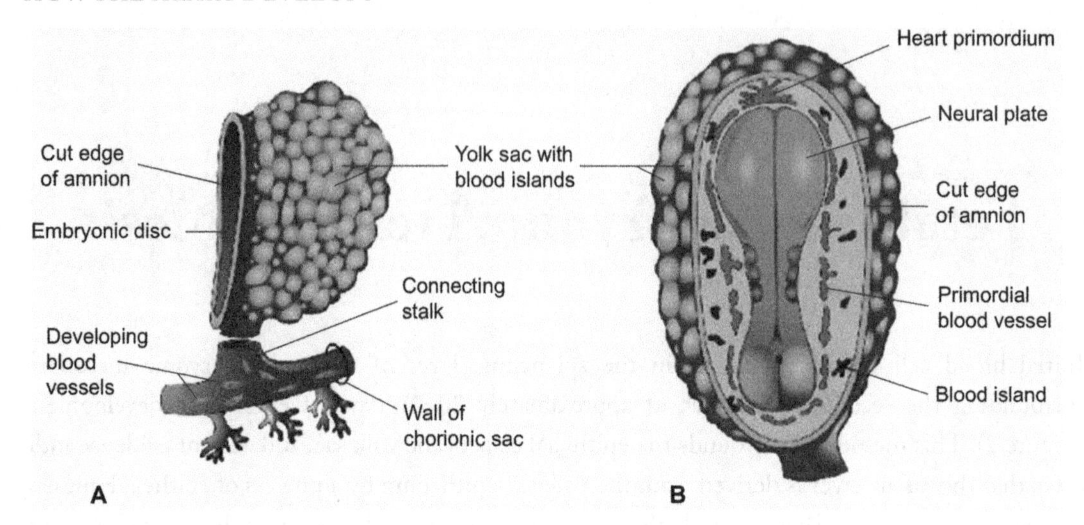


**FIGURE 2:** Vasculogenesis at ~18 days of development. Reproduced from Moore and Persaud. Permission pending.

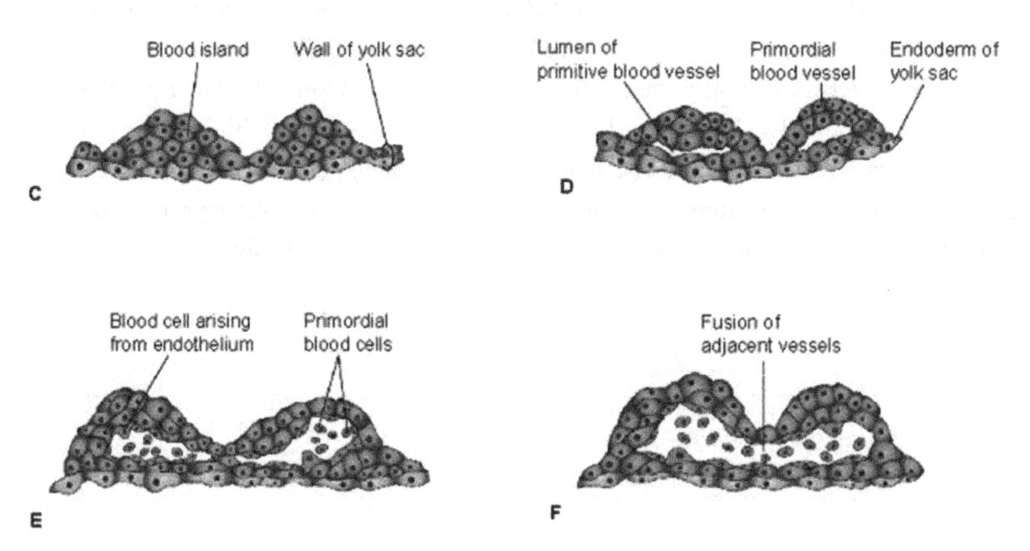


**FIGURE 3:** Blood vessel formation in yolk sac. Permission pending.

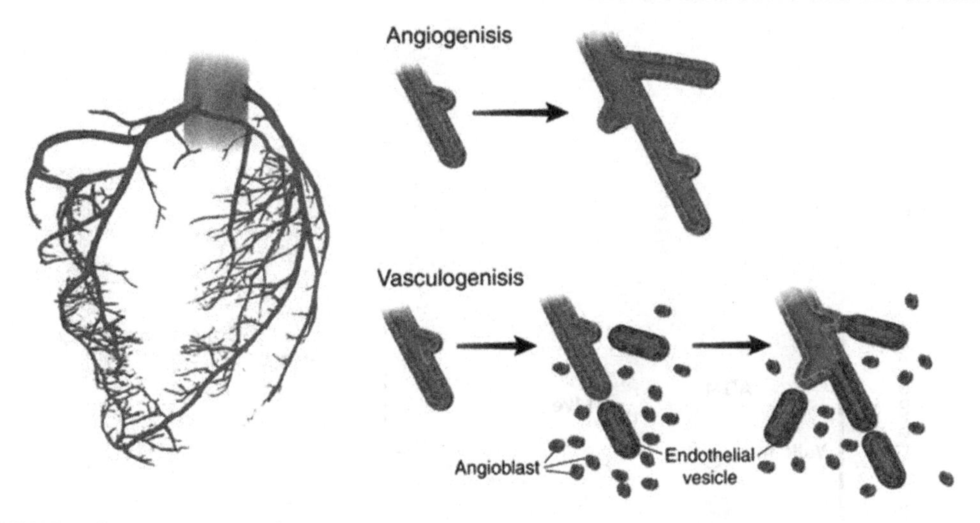


FIGURE 4: Coronary vessels form by vasculogenesis. Permission pending.

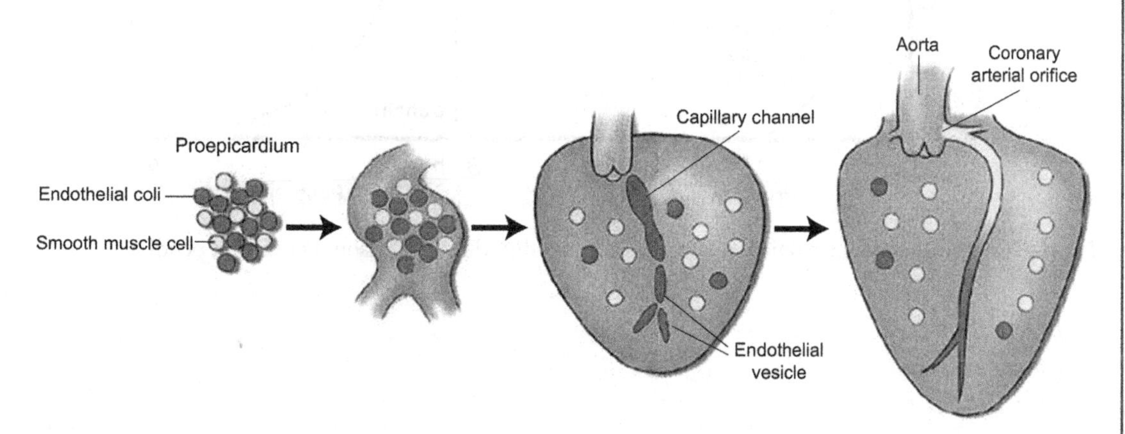


FIGURE 5: Vasculogenesis of coronary arteries. Reproduced from Mikawa and Fischman. Permission pending.

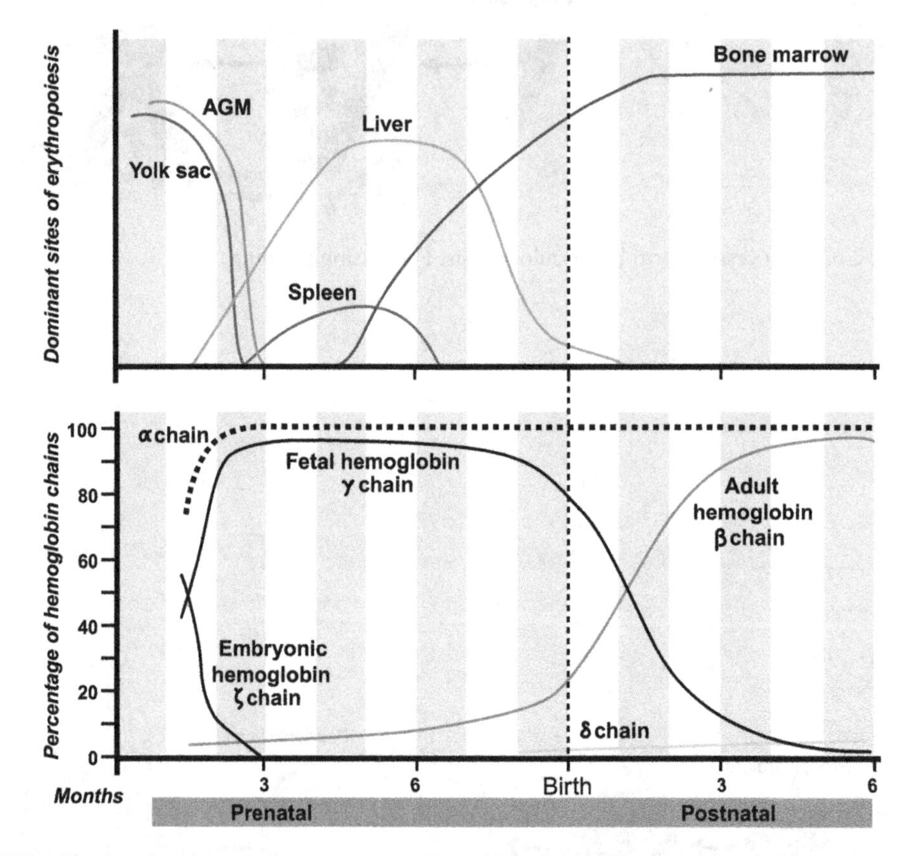


FIGURE 6: Fetal and embryonic hematopoiesis. Reproduced from Carlson. Permission pending.

# Formation of the Precardiac Mesoderm and Fate Mapping During Gastrulation

Fate mapping studies indicate that cells giving rise to the precardiac mesoderm arise in the cranial half of the epiblast (Figure 7). In general, cells forming the outflow track of the heart (bulbus cordis or conus segment) pass through the streak more cranially than those cells forming the inflow track (sinus venosus and atria). Epiblast cells giving rise to the ventricles pass through the streak in its midregion. Once through the streak, the precardiac cells migrate as part of the splanchnic mesoderm, forming a crescent-shaped region of tissue in the anterior half of the embryo termed the *precardiac mesoderm*. This mesoderm is part of the flat-shaped trilaminar disc. Precardiac mesoderm becomes the tubular heart during the lateral and cranial folding of the embryo (Figure 8; Movie 2). The first step in heart formation is the appearance of two parallel endocardial tubes, one on the right and the other on the left side of the embryo in its rostral splanchnic mesoderm (Figure 9). Again, these form by vasculogenesis.

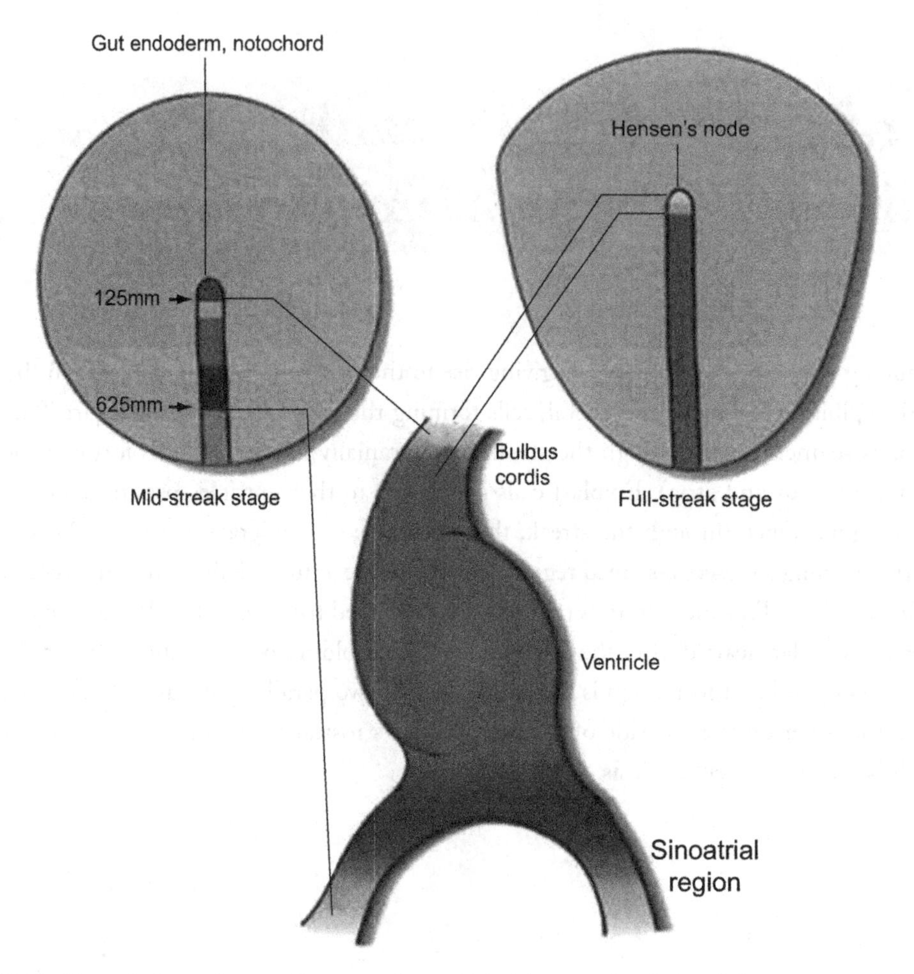


**FIGURE 7:** Gastrulation of precardiac mesoderm. Primitive streak origin of the rostrocaudal subdivisions of the heart in the avian blastoderm. The heart precursors are mapped to a segment of the primitive streak between 125 and 625 mm caudal to the Hensen's mode in the midstreak embryo. Cells that are allocated to the three major heart segments (bulbus cordis, ventricle, and the sinoatrial region) are mapped in the corresponding rostrocaudal order in the primitive streak as in the heart. The Hensen's node does not contribute any cells to the heart. Ingression of the heart precursors finishes by the full-streak stage and there is no further contribution by the primitive streak cells to the heart. Reproduced from Garcia-Martinez and Schoenwolf (1993). Permission pending.

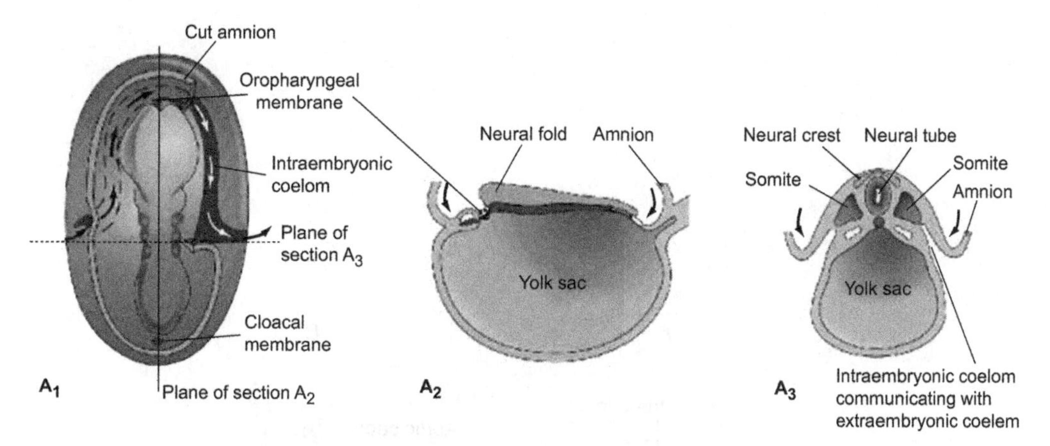


**FIGURE 8:** Cranial, caudal and lateral body folds. Reproduced from Moore and Persaud. Permission pending.

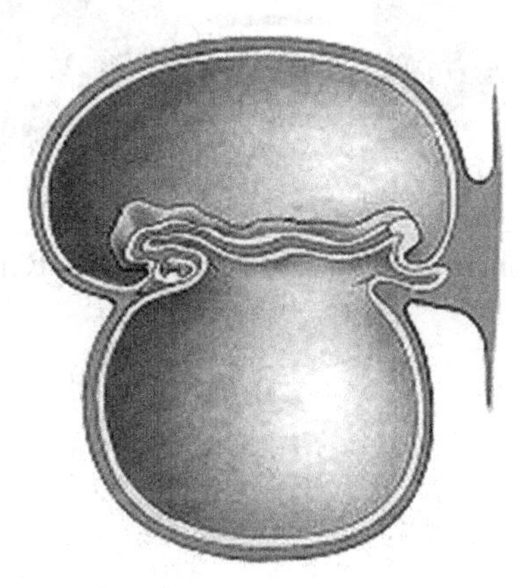

**MOVIE 2:** Cranial and caudal folding (see http://www.morganclaypool.com/userimages/ContentEditor/1245101502851/CranialCaudalFolding.mov).

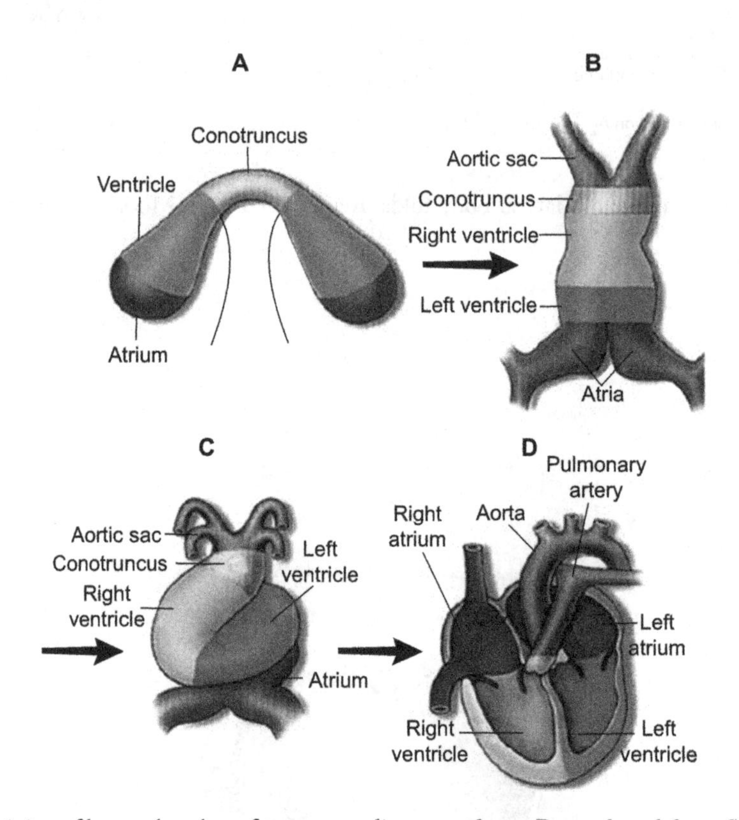


FIGURE 9: Origins of heart chambers from precardiac mesoderm. Reproduced from Srivastava (1999). Permission pending.

# Tubular Heart

Upon ventral flexure of the head fold and ventral convergence of the lateral body folds, the paired endocardial tubes (Figure 10) fuse near the midline, forming a single endocardial tube surrounded by myocardium (Figure 11). These two cell layers, the myocardium and the endocardium, are separated by an acellular hyaluronate-rich, connective tissue matrix (cardiac jelly). The third cell layer (the epicardium) forms by migration onto the myocardium of epithelial and mesenchymal cells derived from the proepicardial organ (or simply the proepicardium) situated on the dorsal body wall of the thorax adjacent to the sinus venosus and just cranial to the future liver (Figure 12). The proepicardium contains all cellular precursors for the coronary arteries (coronary endothelia, smooth muscle and connective tissue), interstitial connective tissue of the myocardium, and the epithelium and connective tissue of the pericardium. In the chicken and mouse, the proepicardium derives from splanchnic mesoderm just caudal to the cardiogenic mesoderm (Figure 13). It is assumed that the same is true for man.

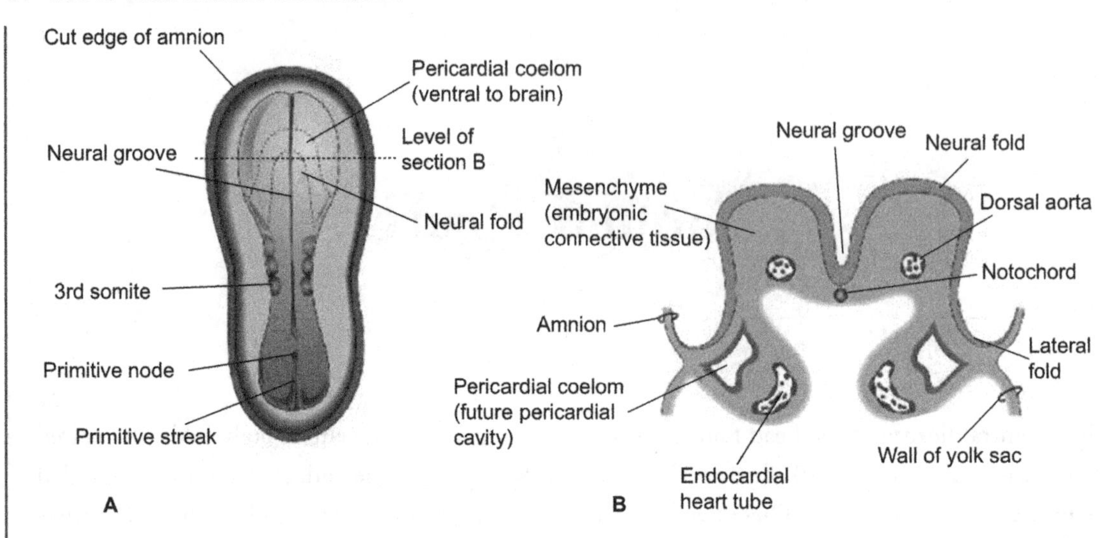


FIGURE 10: Dorsal view of 20-day embryo. Permission pending.

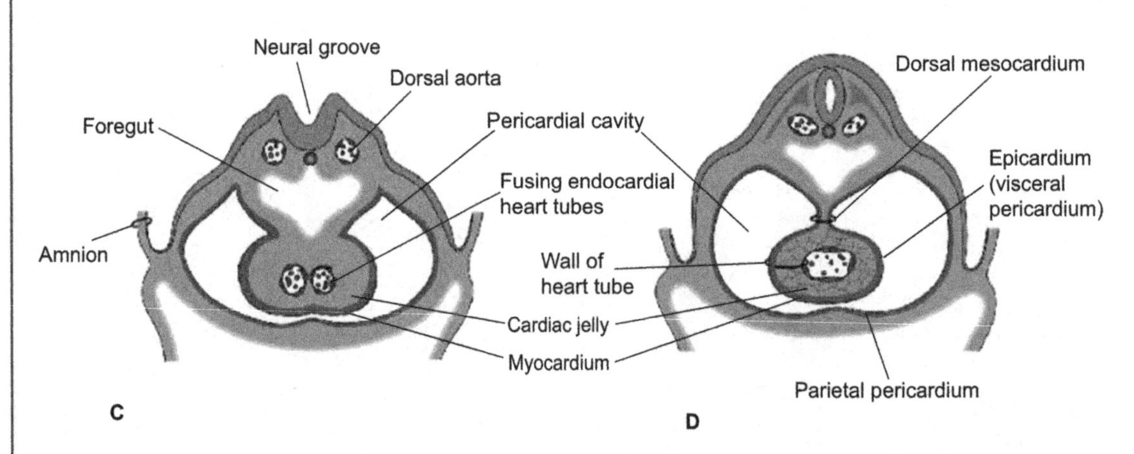


FIGURE 11: Transverse sections of ~21- and 22-day embryos. Reproduced from Moore and Persaud. Permission pending.

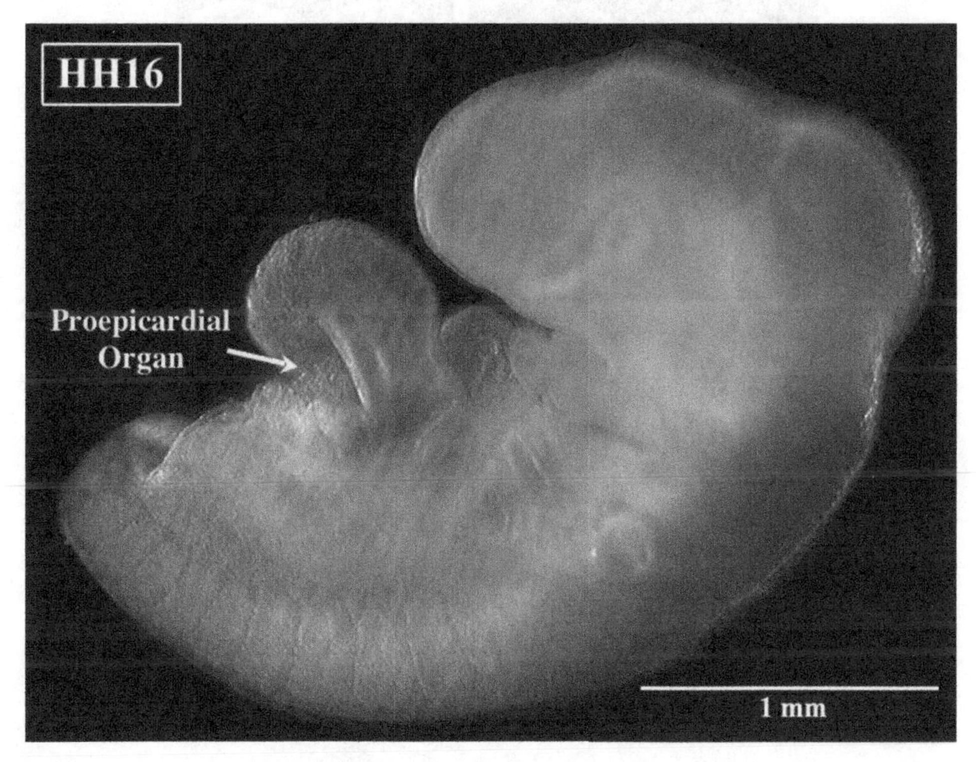


FIGURE 12: Proepicardium. Permission pending.

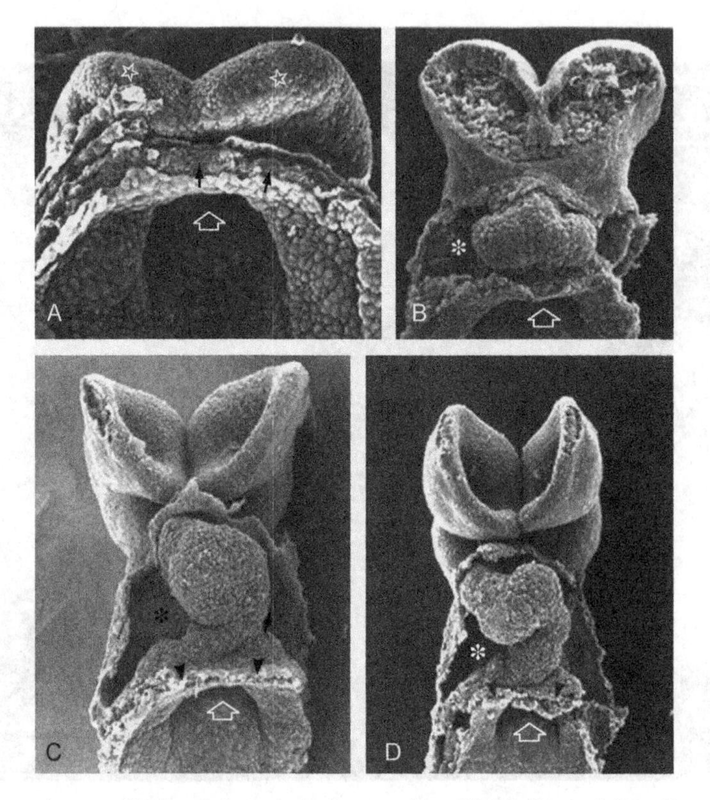


**FIGURE 13:** Mouse heart tube formation equivalent to days 19, 20, 21, and 22 of human development. Reproduced from Sadler. Permission pending.

# Cardiac Looping

The tubular heart is initially composed of an inflow segment—the sinus venosus and presumptive atria (also termed the *sinoatrial segment*), a ventricular segment, and an outflow segment or bulbus cordis (also termed the *conus segment*) (Figure 14, Movie 3). Blood flows in a caudal to cranial direction, since veins from the yolk sac (vitelline vein), placenta (umbilical vein), and body of the embryo (cardinal veins) enter the heart at the sinus venosus. Between days 21 and 28 of embryonic development (E21–28), the inflow tract becomes repositioned dorsally (posteriorally) and cranially to the ventricular and outflow segments by a process termed *cardiac looping*. Basically, there is a dramatic enlargement of the ventricular segment (Figure 15), which moves dextrocaudally and becomes demarcated into right and left ventricular chambers, and at the same time, the inflow region expands cranially and dorsally. The end result is a repositioning of the inflow segment, with the great veins entering the heart near its dorsal, cranial end (Movies 4 and 5).

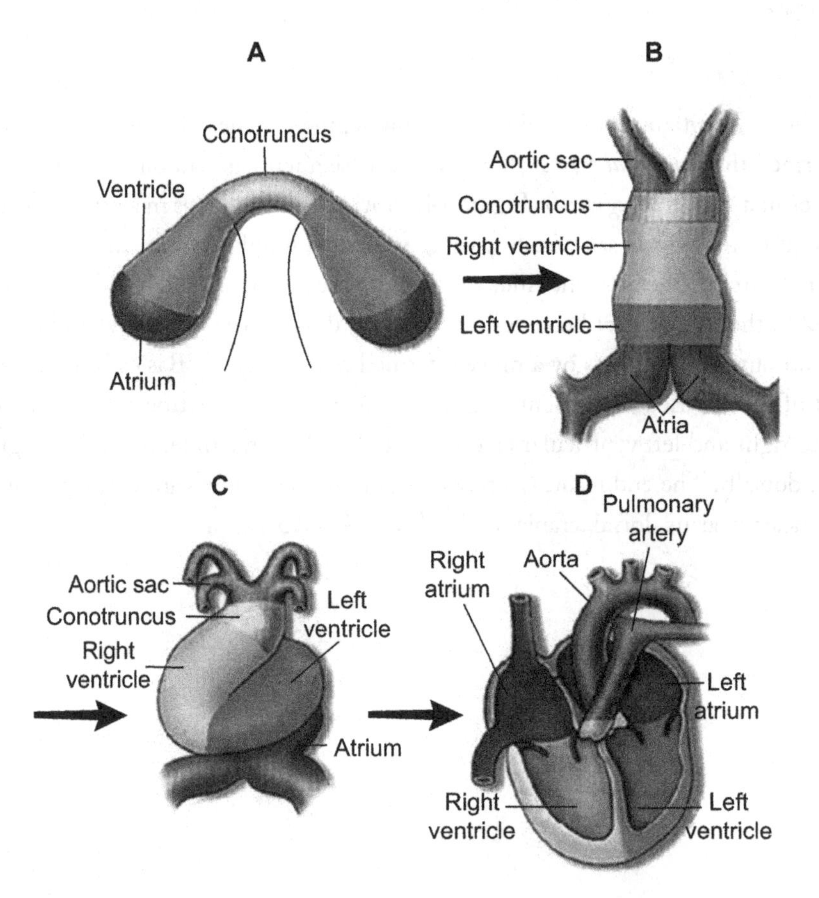


**FIGURE 14:** Origins of heart chambers from precardiac mesoderm. Reproduced from Srivastava (1999). Permission pending.

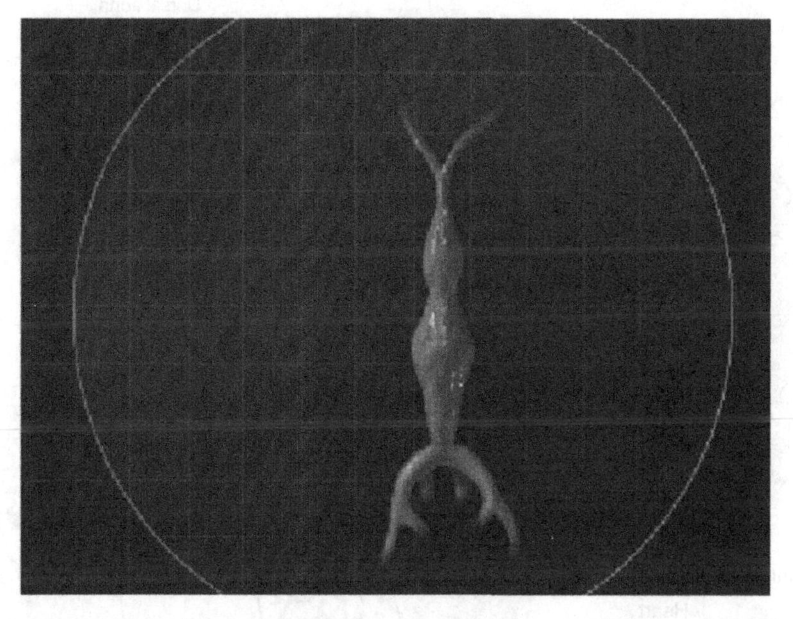

MOVIE 3: Cardiac looping 1 (see http://www.morganclaypool.com/userimages/ContentEditor/1245101424583/CardiacLooping1.MOV).

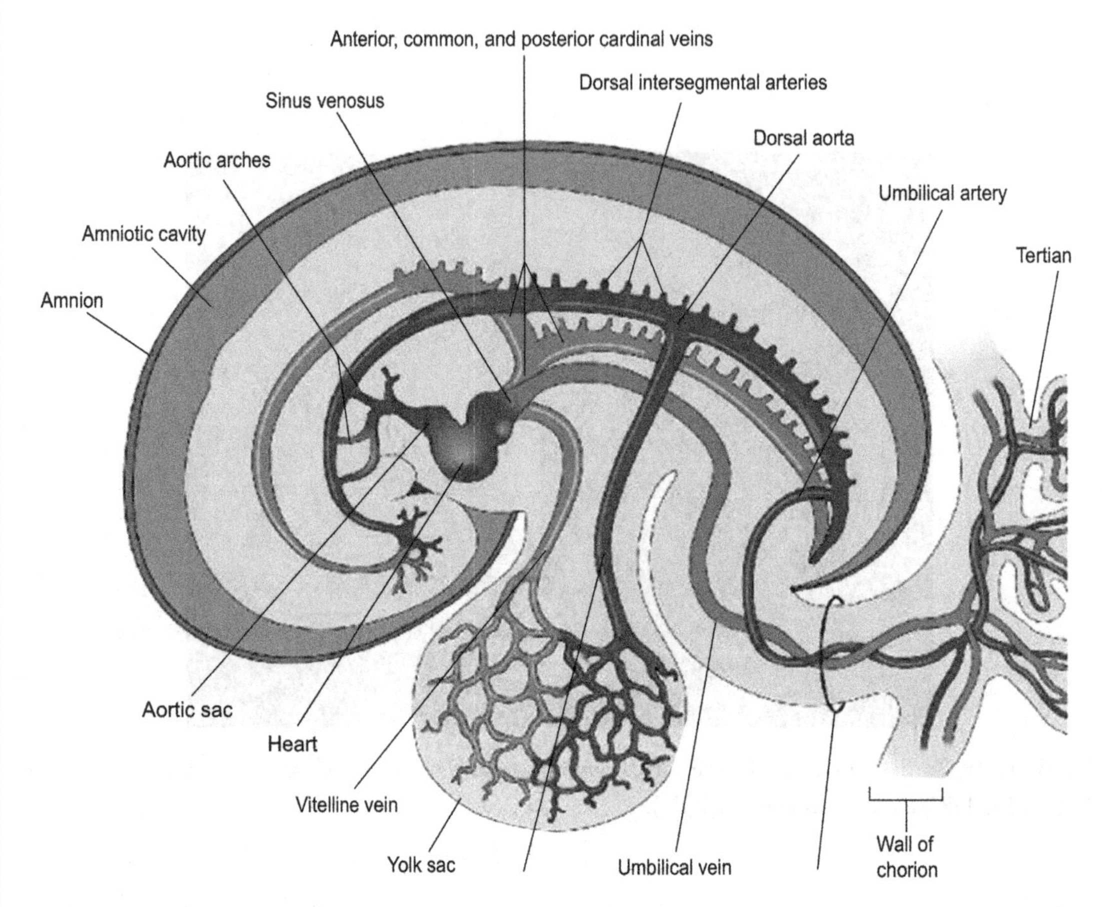


**FIGURE 15:** Circulatory system at E21. Reproduced from Moore and Persaud. Permission pending.

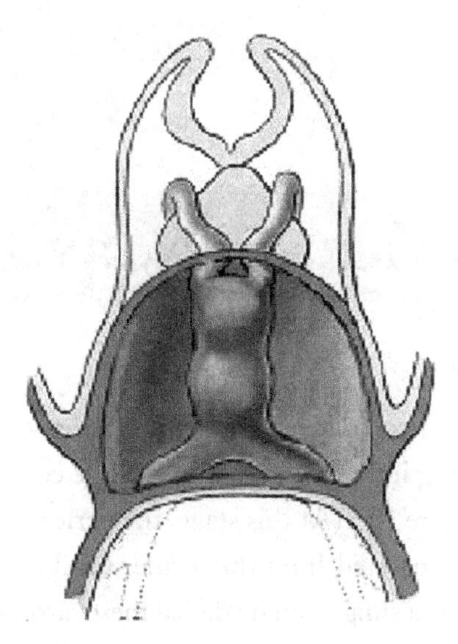

**MOVIE 4:** Cardiac looping 2 (see http://www.morganclaypool.com/userimages/ContentEditor/1245101468405/CardiacLooping2.MOV).

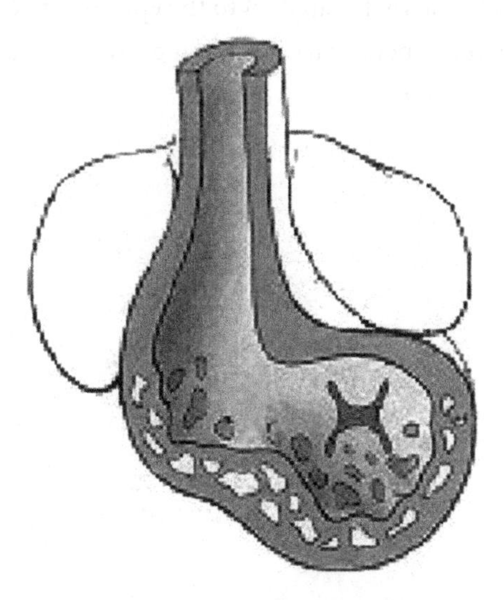

**MOVIE 5:** Rotation and realignment after looping (see http://www.morganclaypool.com/userimages/ContentEditor/1244860996189/RotationandRealignment.mov).

# Pericardial Cavity

When the two endocardial tubes fuse to form the tubular heart, the coelomic cavities beside each endocardial tube also fuse resulting in the formation of a single coelomic cavity enclosing the heart, termed the *pericardial cavity* (Figure 16). (At this stage, the pericardial cavity is identical to the thoracic cavity for the latter is not separated from the former until after lung formation occurs.) The heart is suspended in this cavity as a sling from its dorsal mesocardium connecting the tubular heart to the dorsal body wall. Any cellular elements that migrate to the heart, e.g., the sinus venosus and the aorticopulmonary trunk must pass through this mesocardium. Most of the dorsal mesocardium degenerates soon after it forms, except at the cranial and caudal ends, where the heart tube abuts the dorsal body wall. Thus, along most of its length, the heart hangs as a hammock in the pericardial cavity; the right and left sides of the cavity are in full confluence, dorsally and ventrally. As the heart loops this connection between right and left sides of the pericardial cavity is retained; in the adult this connection on the dorsal side of the heart is termed the *transverse sinus of the pericardium*. The portion of the pericardial membrane that is applied to the epicardium of the heart is termed the *visceral pericardium*. The portion of the pericardial membrane that presses against the outer pericardial wall is termed the *parietal pericardium*.

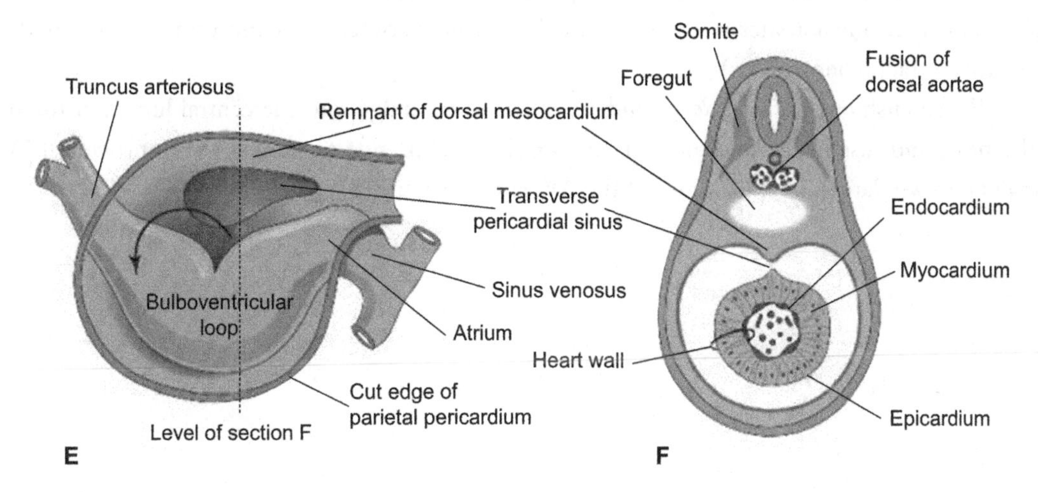


**FIGURE 16:** Parasagittal and transverse sections at E28: transverse pericardial sinus. Reproduced from Moore and Persaud. Permission pending.

# Endocardial Cushions

Septation of the atria from the ventricles and division of the conotruncus into aortic and pulmonary trunks occur by the convergence of endocardial prominences termed the *endocardial cushions* (Figure 17). These structures are derived from an epithelial–mesenchymal transformation of endocardial epithelium at two major sites in the heart: the borderline between atria and ventricles and within the outflow tract (conotruncus).

Paired cushions (termed *dorsal* and *ventral cushions*) bulge into the central lumen of the tubular heart and fuse in the midline, but retain right and left side openings (atrioventricular [AV] canals) that will later be the locations of the AV valves (Figure 18).

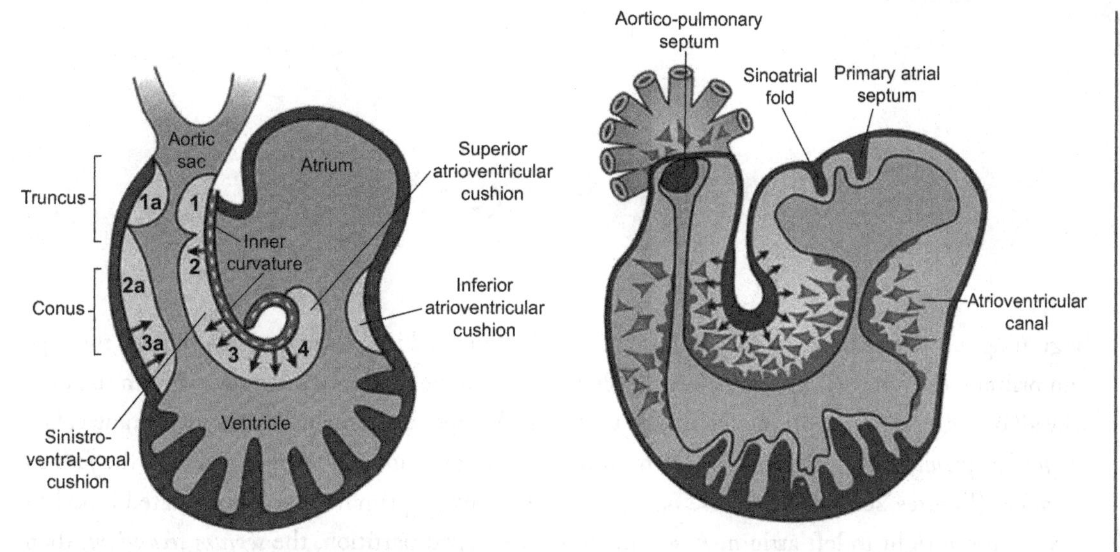


**FIGURE 17:** Endocardial cushions and remodeling of lesser curvature. Segmental interaction during remodeling of the inner curvature. (*Left*) The relative positions (1–5) of the endocardial cushion swellings are shown within the looping heart tube (approximately HH stage 24). The outlow tract of the heart consists of a proximal region defined here as the conus and a more distal part defined as the truncus arteriosus. The myocardium of the heart's inner curvature is completely lined with cushion tissue (2–4) by the fusion of the sinistroventroconal cushion (SVCC) and the superior atrioventricular cushion (SAVC). The inner curvature myocardium (dashed line) is removed (arrows) during remodeling to allow proper septal alignments. The inner curvature myocardium is the only region of the heart that comes into close interaction with cushion mesenchyme derived from two separate heart segments (SAVC and SVCC). (*Right*) The myocardial cells invade (arrows) the cellularized cushions that line the inner curvature of the heart tube. Reproduced from Markwald. Permission pending.

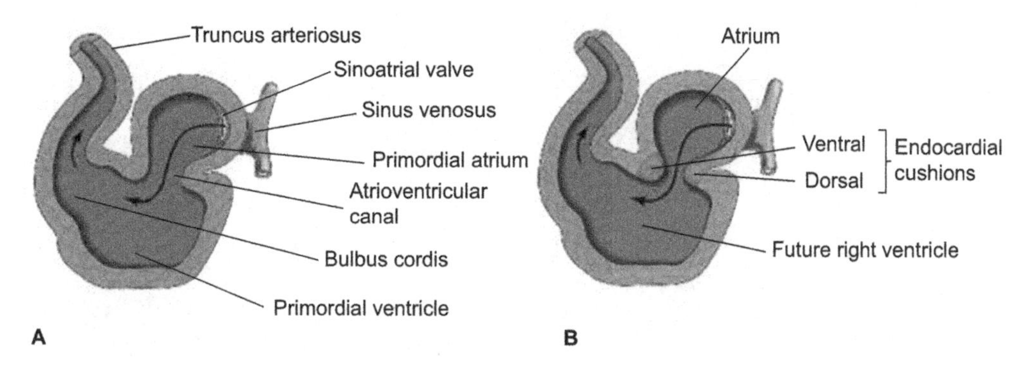


**FIGURE 18:** Modification of AV cushions: fourth to fifth week. Reproduced from Moore and Persaud. Permission pending.

# Atrial Septation

Beginning in the fourth week, the primordial atrium is divided into right and left atria by the septum primum (Figure 19). This is a crescent-shaped membrane that grows downward from the dorsal wall of the atrium toward the fusing AV cushions. An opening remains in this membrane called the *formen primum*. Perforations then form in the septum primum and these coalesce as the *formen secundum* (Figures 20 and 21). These openings in the septum primum allow oxygenated blood to flow from the right to left atrium. A second crescent-shaped partition, the *septum secundum*, then descends from the cranial side of the right atrium (just to the right side of the septum primum) and toward the caudal, dorsal end of the atrium (Figure 22). However, it leaves an inferior flap open to the foramen secundum, which will later become the valve of the *foramen ovale*, as the foramen secundum is eventually termed. With closure of the foramen primum (Figure 23), the foramen ovale is the single channel permitting oxygenated blood from the umbilical vein to reach the left atrium, bypassing the right ventricle (Figure 24).

The overall process of atrial septation are presented in Movies 6 and 7.

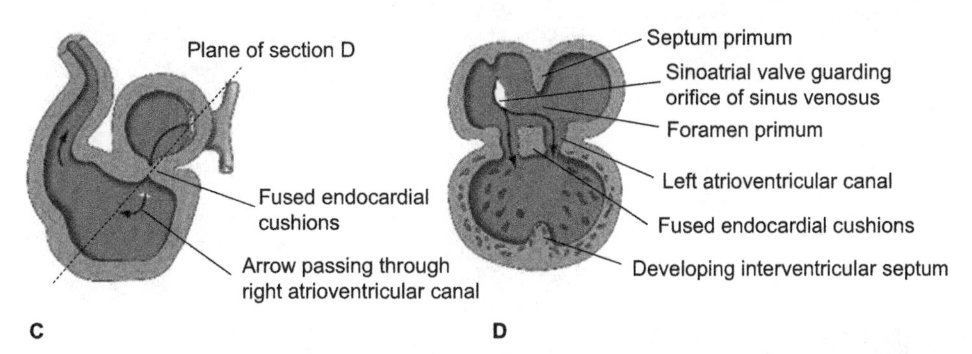


**FIGURE 19:** Division of AV canal and formation of septum primum: fifth week of development. Reproduced from Moore and Persaud. Permission pending.

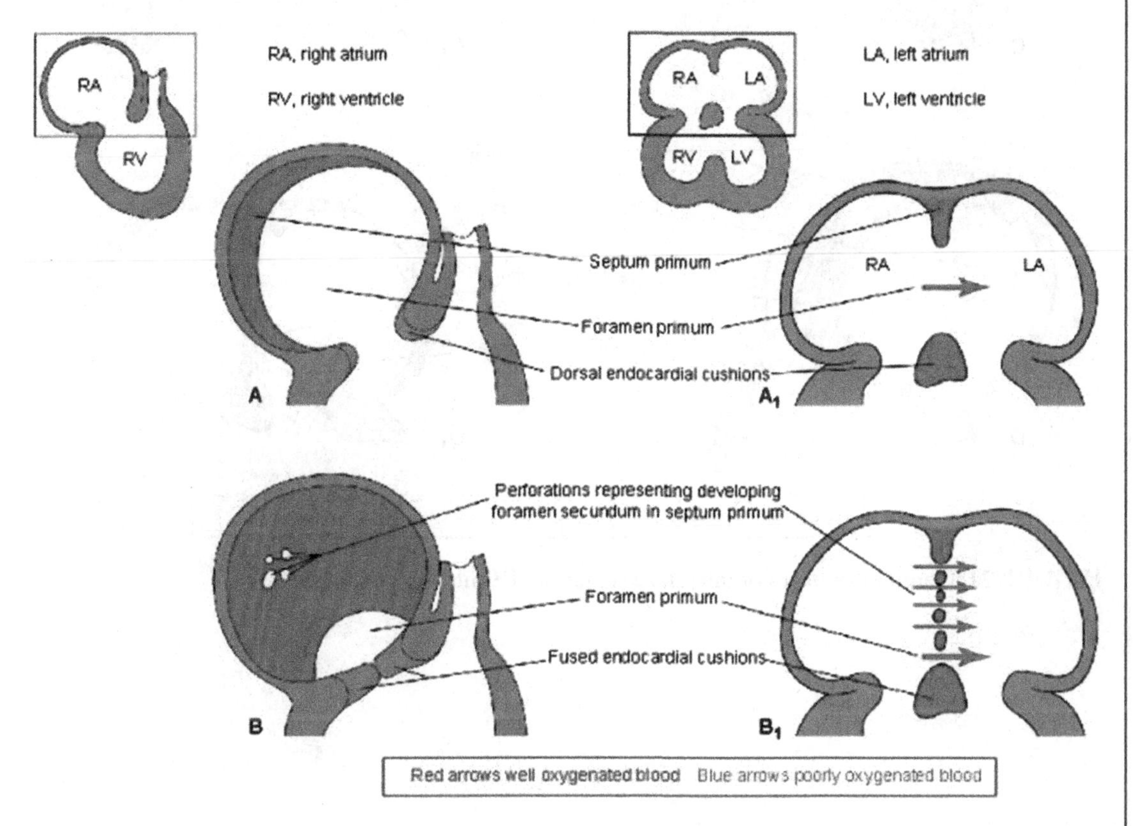


**FIGURE 20:** Progressive stages of interatrial septation. Permission pending.

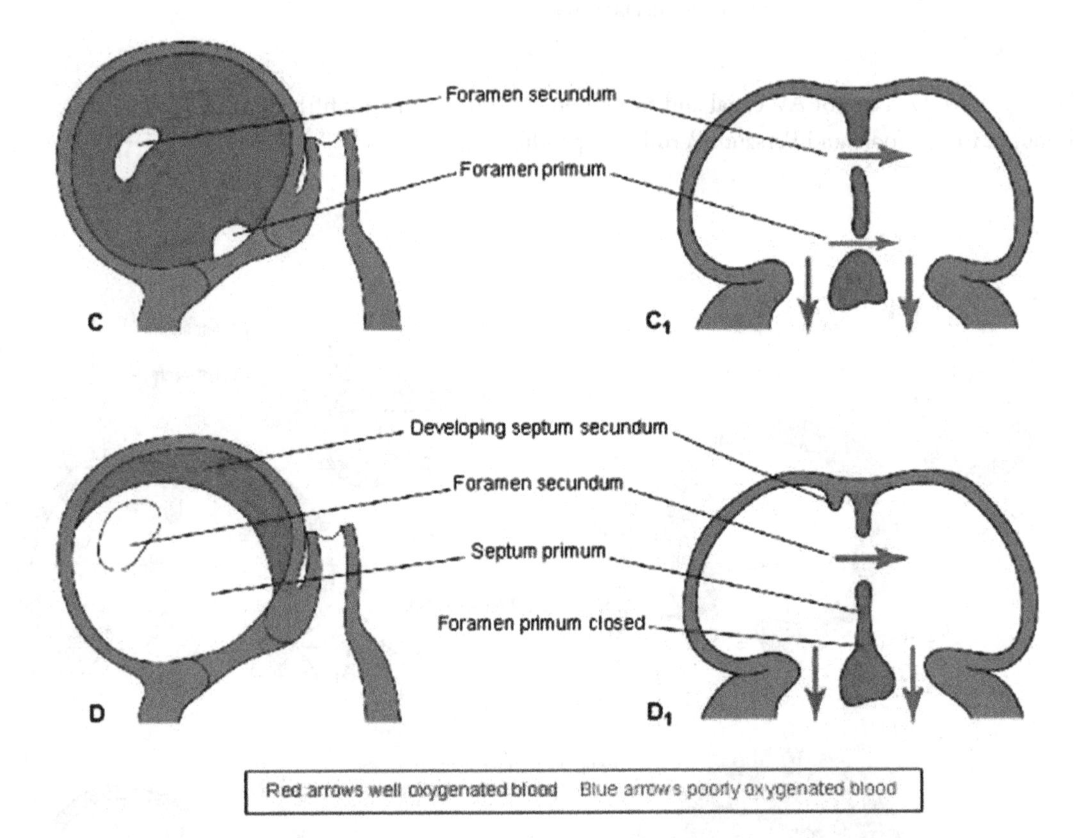


**FIGURE 21:** Continuing stages of interatrial septation. Permission pending.

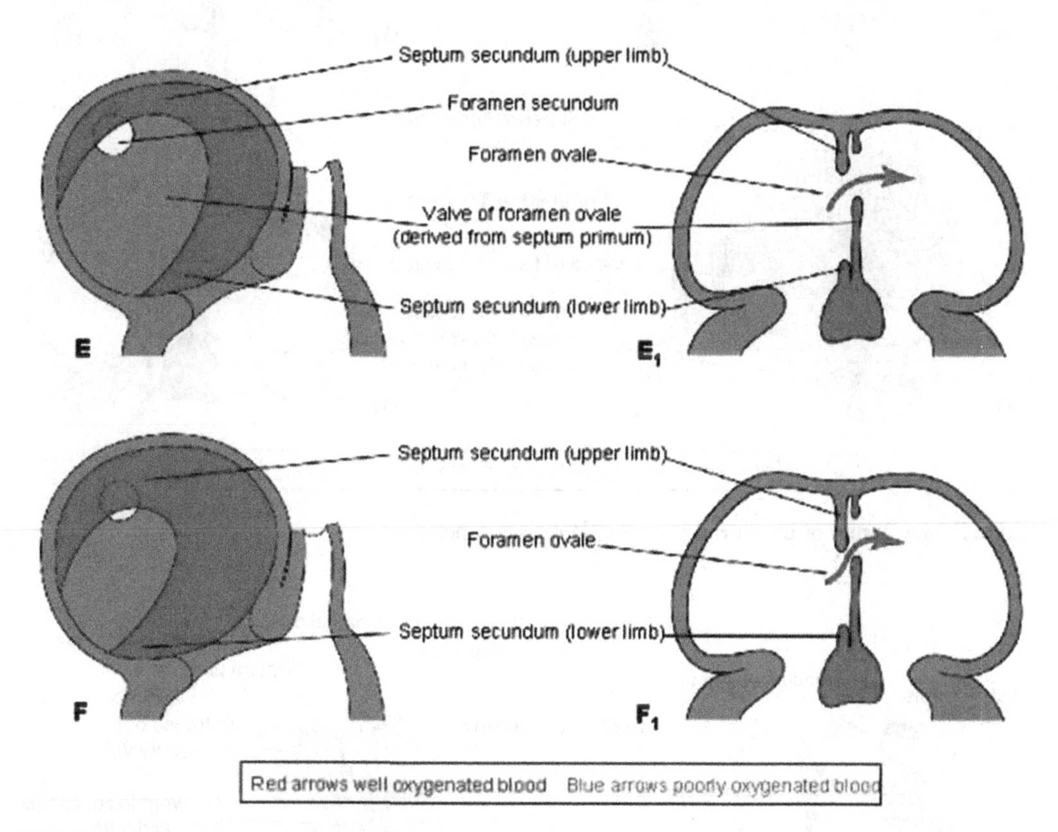


FIGURE 22: Further stages of interatrial septation. Permission pending.

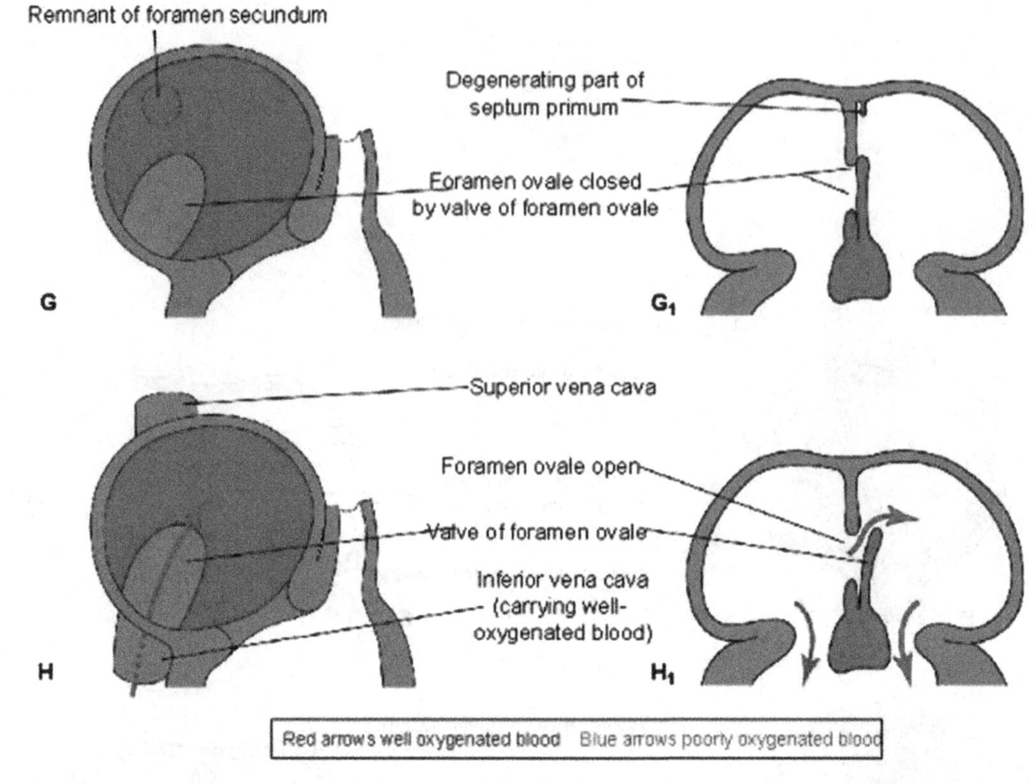


FIGURE 23: Closure of interatrial foramen. Permission pending.

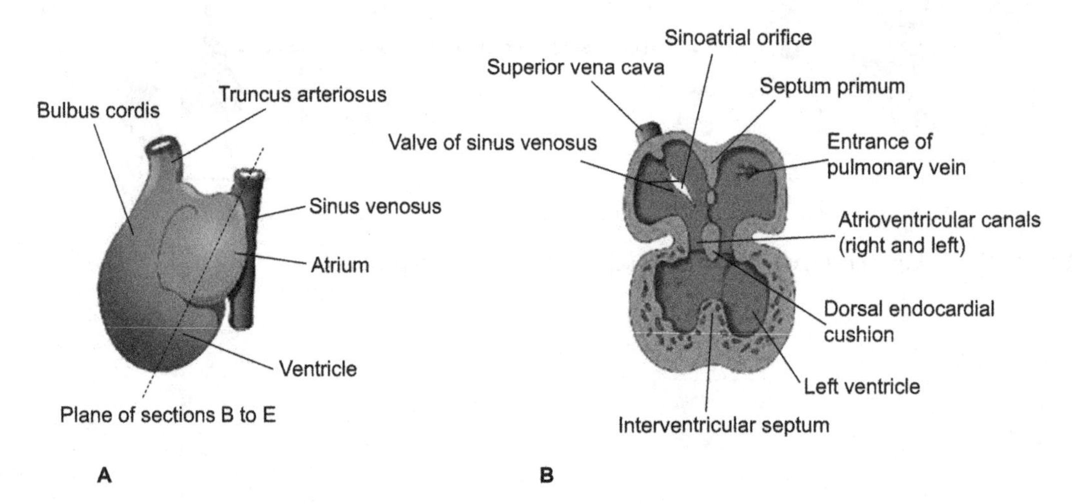


FIGURE 24: Fourth week: septum primum, IV septum, and dorsal AV endocardial cushion. Reproduced from Moore and Persaud. Permission pending.

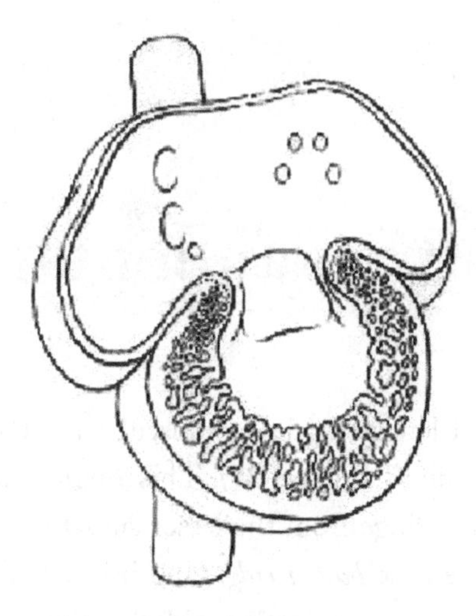

MOVIE 6: Atrial septation 1 (see http://www.morganclaypool.com/userimages/ContentEditor/1244860849114/AtrialSeptation.MOV).

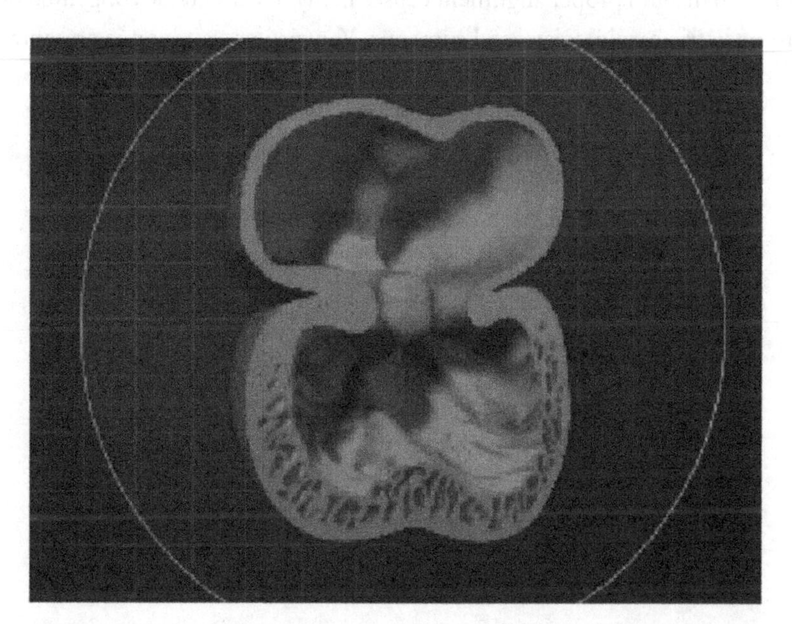

MOVIE 7: Atrial septation 2 (see http://www.morganclaypool.com/userimages/ContentEditor/1245427857107/AtrialSeptation2.MOV).

# Ventricular Septation

At the junction of the right and left ventricles, a muscular ridge elevates from the inferior wall of the endocardium and myocardium (Figure 25). This ridge is directed toward the fusing AV cushions and toward a descending ridge from the lesser curvature of the heart (cranial end of the left ventricle). This ridge is termed the *bulbar ridge*, and it eventually fuses with the AV cushions. A membrane connects the AV cushions and bulbar ridge to the muscular ridge: this membranous connection will become the membranous portion of the interventricular (IV) septum; the inferior muscular ridge will become the muscular portion of the IV septum. However, an opening remains in the IV septum until the seventh week of development, permitting blood flow between right and left ventricular chambers (Figure 26). It is absolutely critical that the IV septum meet the bulbar ridge where it fuses to the AV cushions. Failure of these three elements (IV septum, bulbar ridge, and AV cushions) to meet in proper alignment causes many of the cardiac congenital defects seen by neonatologists, pediatric cardiologists, and surgeons. Ventricular septation is presented in Movie 8.

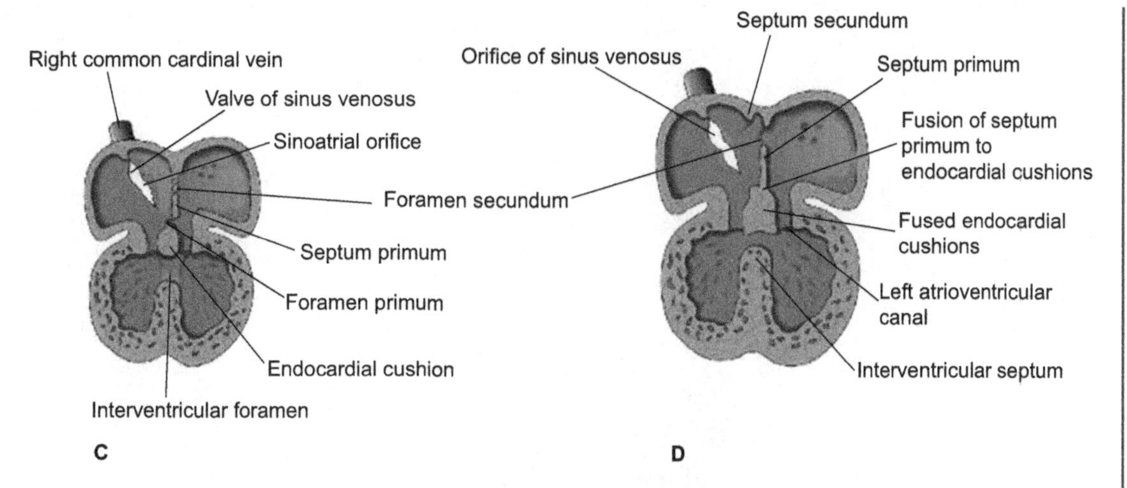


FIGURE 25: Approximately ~E32: septum primum, septum secundum, and primary and secondary interatrial foramina. Reproduced from Moore and Persaud. Permission pending.

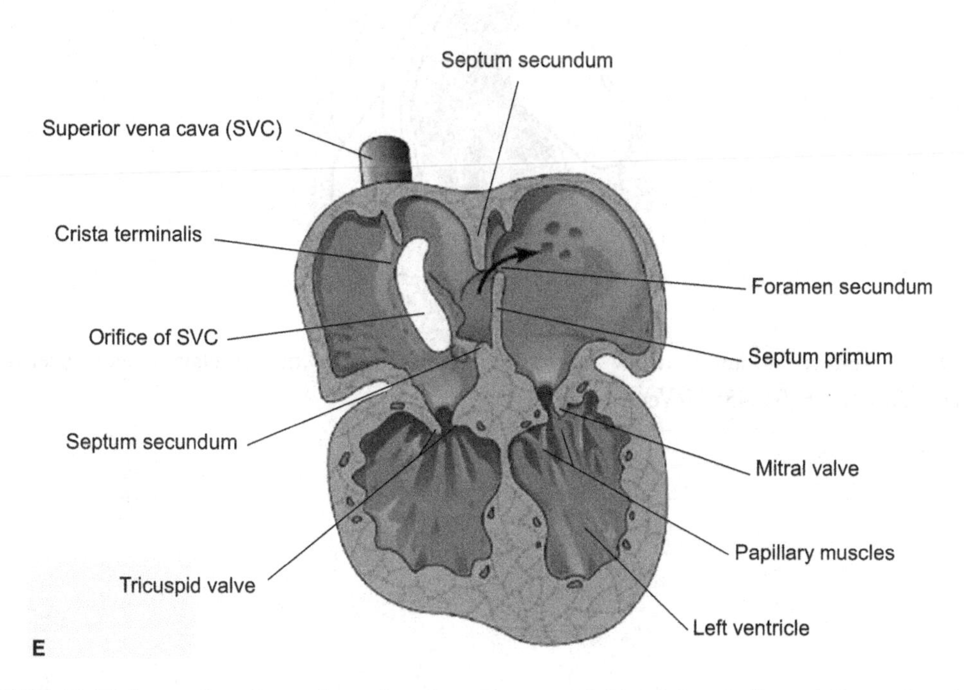


FIGURE 26: Eight-week embryo: four-chambered heart with blood flow indicated. Reproduced from Moore and Persaud. Permission pending.

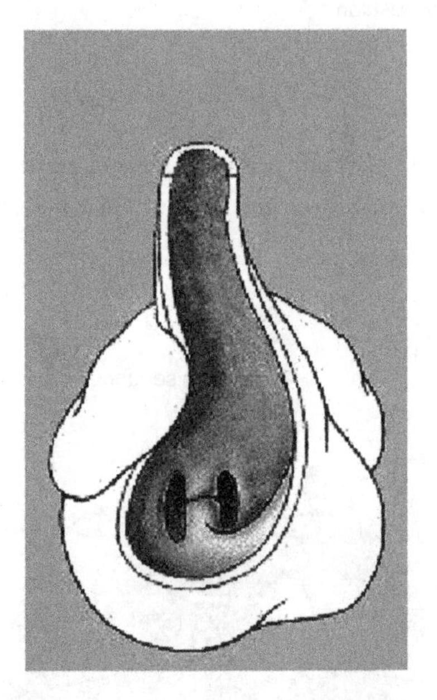

**MOVIE 8:** Ventricular and outflow septation (see http://www.morganclaypool.com/userimages/ContentEditor/1245101545595/Ventricular.mov).

# Partitioning of the Bulbus Cordis and Truncus Arteriosus

During the fifth week of development, two sets of longitudinal ridges form along the bulbus cordis and truncus arteriosus, the *bulbar* and *truncal ridges*, respectively (Figures 27 and 28). This process is also called partitioning of the truncus. Mesenchyme in these ridges is derived from both endocardium (epithelial–mesenchymal transformation) and neural crest that migrated from the pharyngeal arches to the outflow tract of the heart. The bulbar and truncal ridges undergo a 180° spiraling. This results in the formation of a spiral aorticopulmonary septum when the ridges fuse (Figure 29). This septum divides the bulbus cordis and truncus into two arterial channels, the aorta and pulmonary trunk. The bulbus is eventually incorporated into the walls of the definitive ventricles: on the right, the bulbus becomes the conus arteriosus (infundibulum) that flows into the pulmonary trunk. On the left, the bulbus forms the walls of the aortic vestibule, just inferior to the aortic valve. The semilunar valves form just as the partitioning of the truncus is nearly complete.

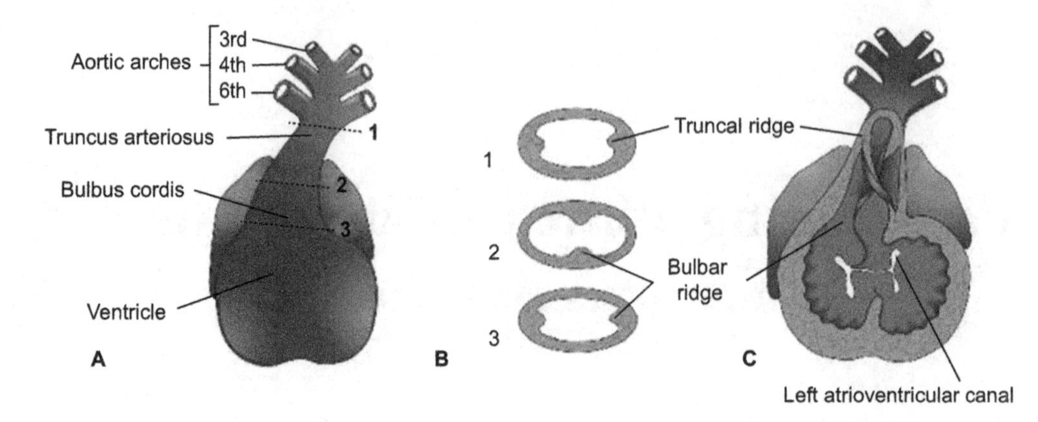


FIGURE 27: Partitioning of the outflow tract: 1, the conotruncal ridges. Reproduced from Moore and Persaud. Permission pending.

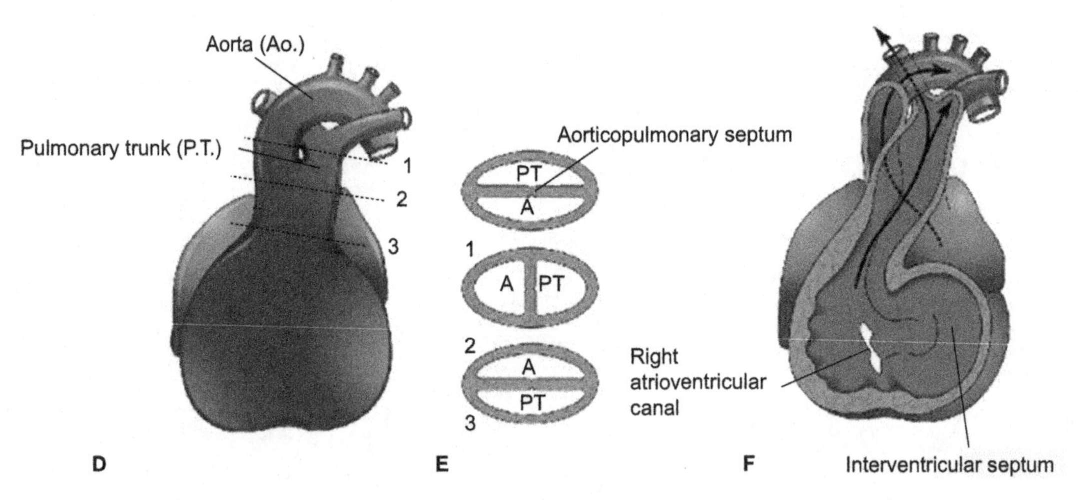


FIGURE 28: Partitioning of the outflow tract: 2. Reproduced from Moore and Persaud. Permission pending.

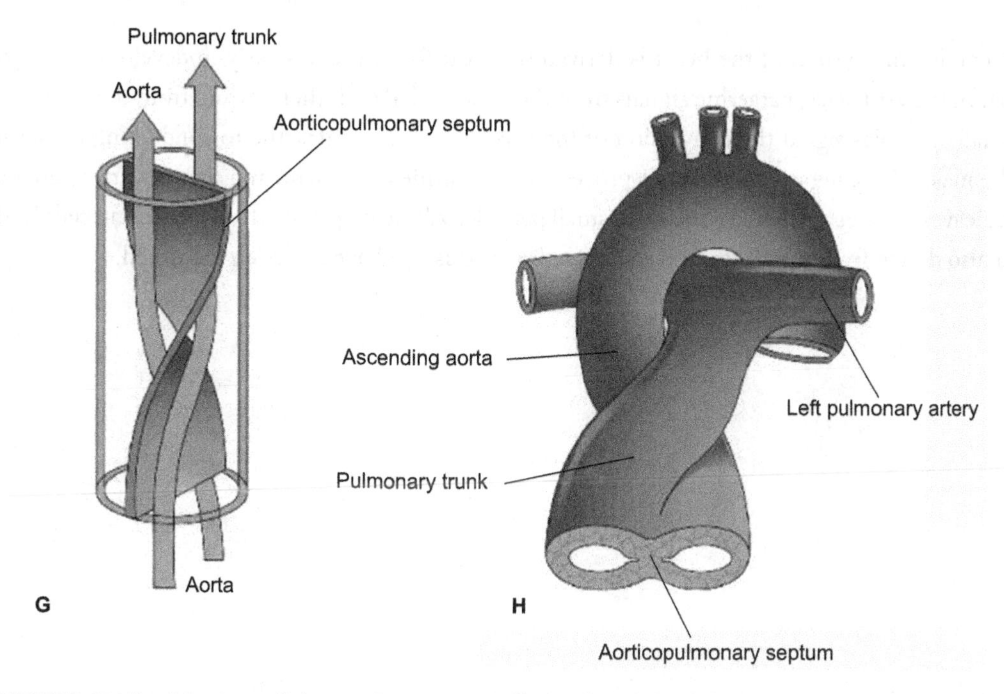


**FIGURE 29:** Partitioning of the outflow tract: 3. Reproduced from Moore and Persaud. Permission pending.

# Conducting System

The conducting system of the heart is derived from cardiac myocytes early in development of the heart. In the ventricles, paracrine signals from the endocardial endothelium and from the endothelia of small arterioles signal the conversion of the myocytes from contractile to conducting cells. One of the major signaling cascades in this process is the peptide endothelin and the converting enzyme that cleaves large endothelin to its active small peptide. While it appears that conducting cells in the atria also derive from myocytes, the signals acting at this site have not been identified.

# Cell Lineages During Heart Development

Using a variety of cell markers, but mainly replication-defective retroviruses, all of the cell lineages in the heart have been identified. They are summarized in the accompanying diagram (Figure 30). A surprising result of such studies was the demonstration that the proepicardium, an anlagen that forms the epicardium of the heart, also gives rise to all of the cells that comprise the coronary arteries and the connective tissues of the atria and ventricles (Figure 31). The cardiac neural crest gives rise to smooth muscle essential for formation of the aortic and pulmonary trunks and components of the outflow (aorticopulmonary) septum (Figure 32).

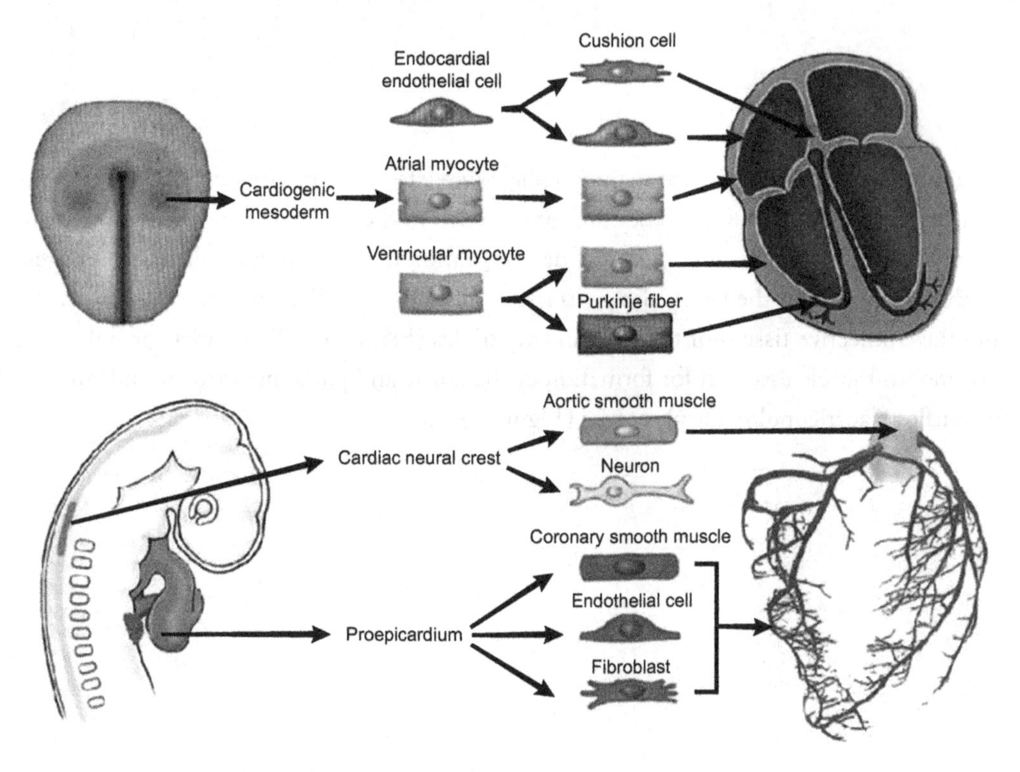


**FIGURE 30:** Summary of cell lineages in the heart. Reproduced from Mikawa and Fischman. Permission pending.

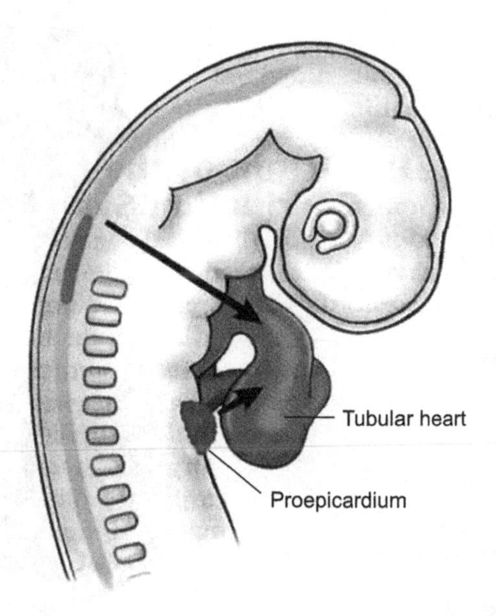


**FIGURE 31:** Proepicardium: formation of the epicardium and coronary arteries. Reproduced from Mikawa and Fischman. Permission pending.

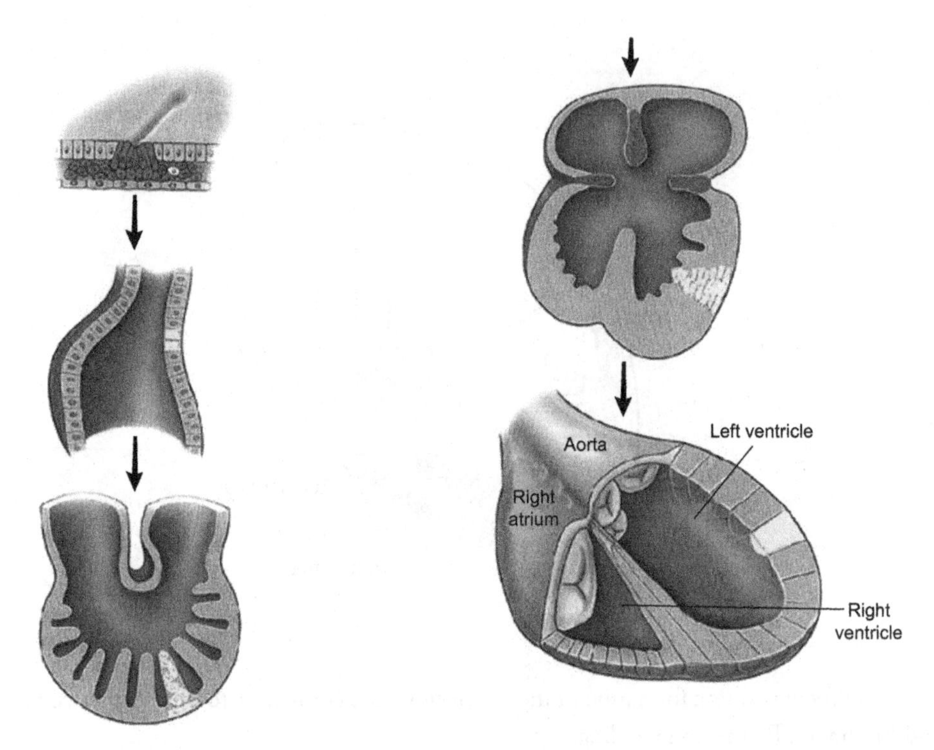


FIGURE 32: Growth of clones in the myocardium. Reproduced from Mikawa and Fischman. Permission pending.

# Circulation at Term and Changes upon Birth

In the fetus, oxygenated blood reaches the baby via the umbilical vein (Figure 33). Most of this blood short-circuits the liver to enter the inferior vena cava via the ductus venosus. After entering the right atrium (Figure 34), much of this blood passes through the foramen ovale and reaches the left atrium bypassing the right ventricle, the pulmonary circulation, and the lungs. This blood is then pumped by the left ventricle into the arch of the aorta to reach the arteries supplying the upper half of the body, especially the head. Blood from the superior vena cava and some of the oxygenated blood from the inferior vena cava passes through the tricuspid valve and enters the right ventricle. This blood then enters the pulmonary artery, where it either enters the lungs or bypasses the lungs via the ductus arteriosus reaching the descending aorta. Deoxygenated blood leaves the fetus via the paired umbilical arteries. This circulation ensures that oxygenated blood reaches the brain, and less oxygenated blood reaches the lungs and other body organs. The lungs receive only that amount of blood sufficient for their growth.

Obviously, at birth, this must change radically since the fetus goes from being an aquatic fetus to an air-breathing newborn baby. At birth, blood from the placenta is cut off and the lungs quickly expand and begin function (Figure 35). The three shunts that permitted blood to bypass the liver and lungs close and cease to function. As soon as the baby is born, the foramen ovale, ductus arteriosus, ductus venosus, and umbilical vessels are no longer needed. A sphincter in the ductus venosus closes and the ductus venosus constricts. This causes a rapid fall in blood pressure in the inferior vena cava and right atrium. Aeration of the lungs causes a dramatic fall in pulmonary vascular resistance, a large increase in pulmonary blood flow and a stretching of the walls of the pulmonary arteries. Associated with an increase in pulmonary blood flow, there is a rise in left atrial pressure and this closes the flap valve overlying the foramen ovale. Blood flow from right to left atria ceases. Because blood pressure in the pulmonary artery falls (decreased pulmonary resistance), there is a reversal of blood flow in the ductus arteriosus; blood flow shifts from the descending aorta to the pulmonary artery. This is accompanied by an increase in left ventricular work and a consequent thickening of the left ventricular wall. The ductus arteriosus rapidly constricts. This appears to be mediated by two major factors: secretion of bradykinin (constricts ductus arteriosus smooth muscle) from the lungs as they initiate function and the fall in local secretion of prostaglandin $E_2$ and prostacyclin $I_2$ (which are needed to keep the ductus arteriosus relaxed). The end result is that the ductus arteriosus constricts and closes completely within 24–48 hours after birth of a normal child.

Some major cardiac birth defects are given in Figures 36–44.

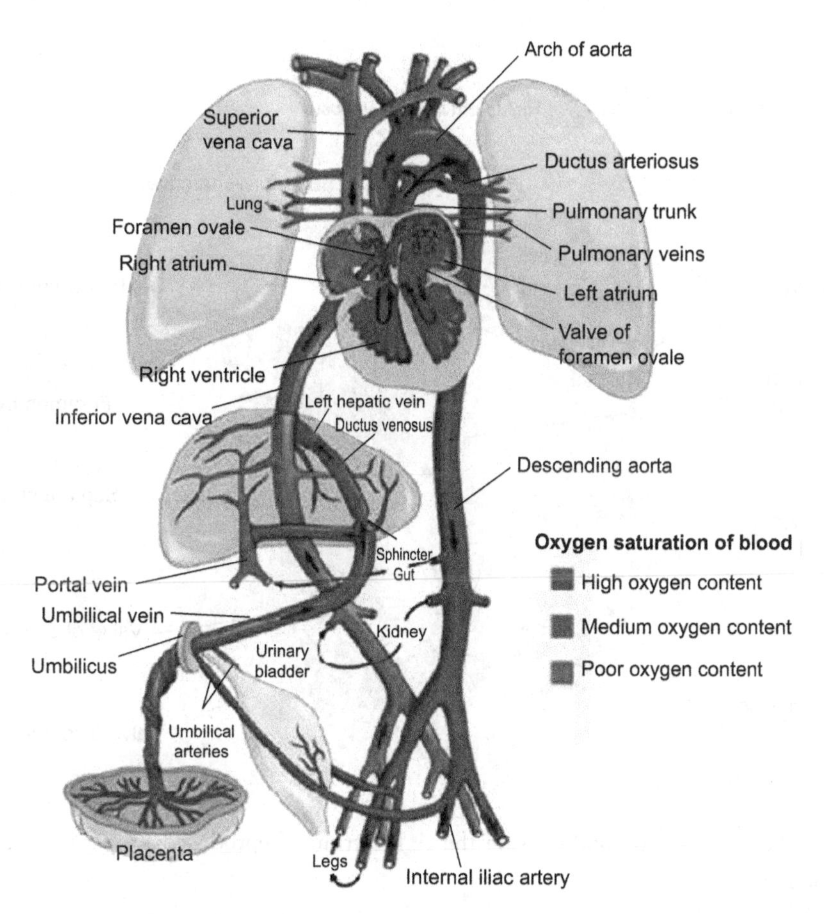


FIGURE 33: Fetal circulation before birth. Reproduced from Moore and Persaud. Permission pending.

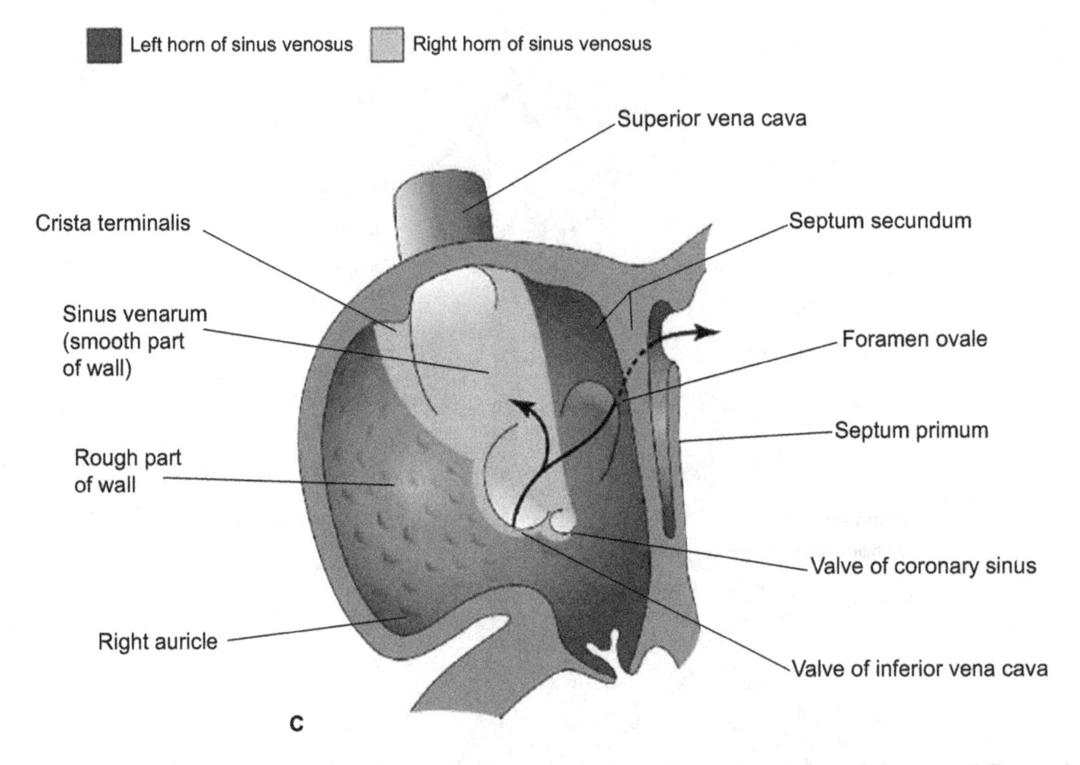


FIGURE 34: Anatomical landmarks inside the right atrium. Reproduced from Moore and Persaud. Permission pending.

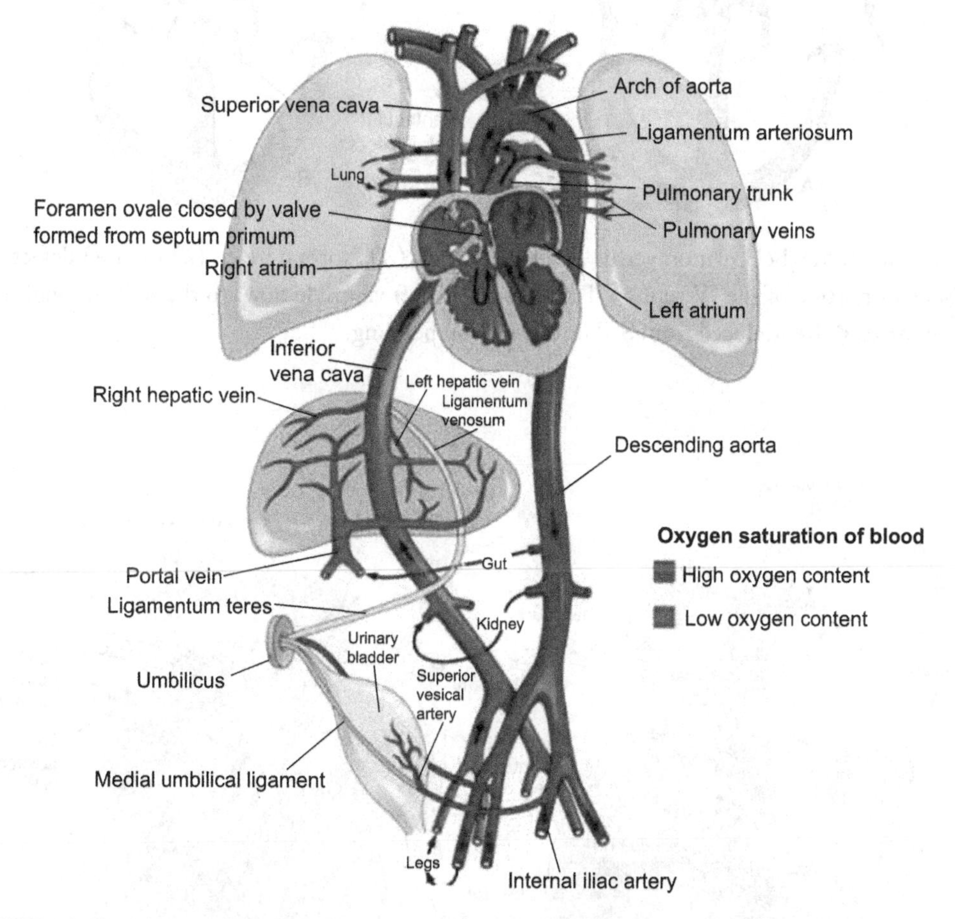

**FIGURE 35:** Circulation after birth. Reproduced from Moore and Persaud. Permission pending.

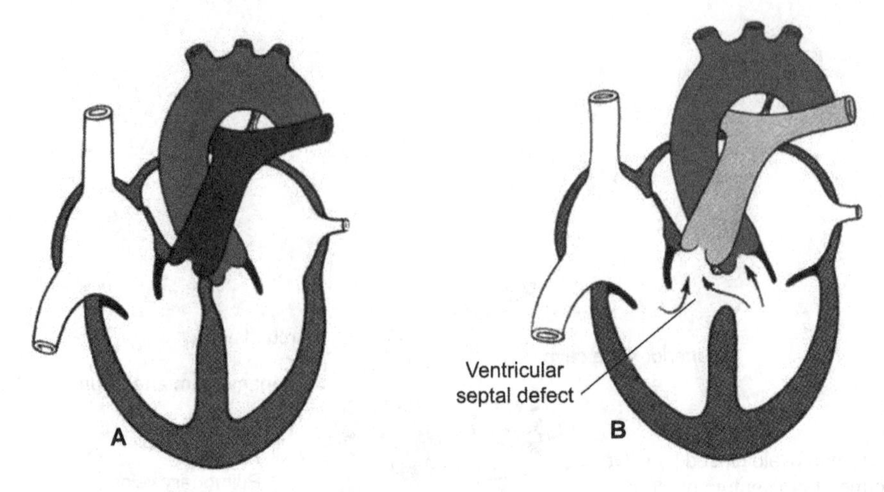


FIGURE 36: Normal circulation versus IV septal defect. (*A*) Normal heart. (*B*) Isolated defect in the membranous portion of the IV septum. Blood from the left ventricle flows to the right through the IV foramen (*arrows*). Reproduced from Sadler. Permission pending.

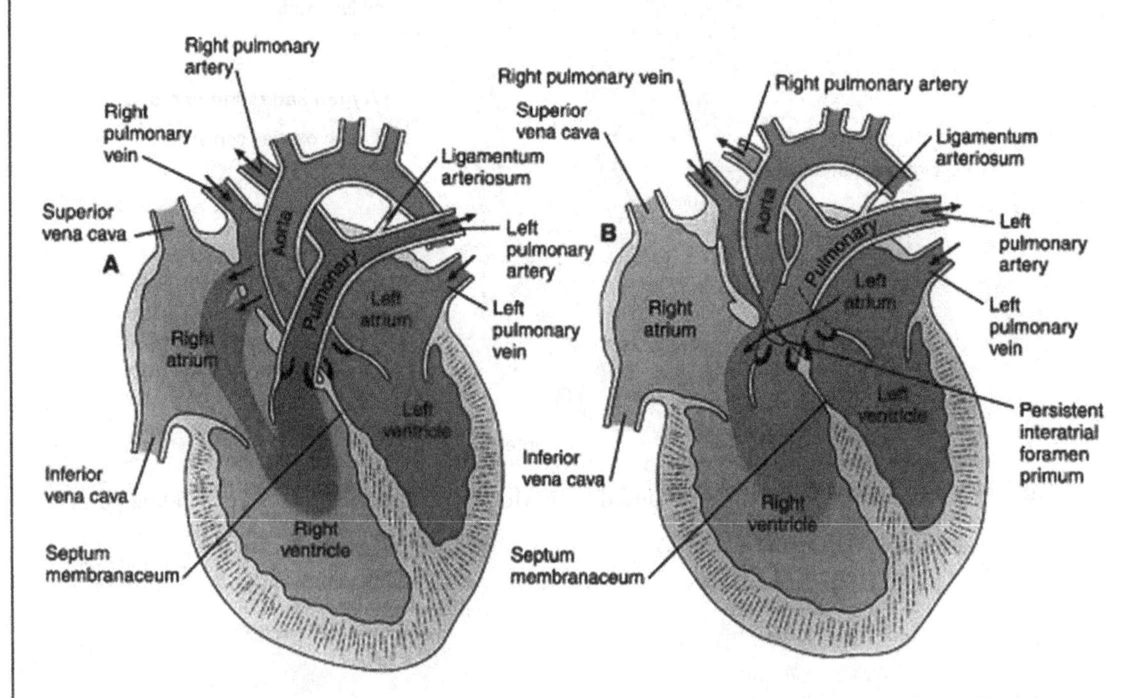


FIGURE 37: Atrial septal defects. High (*A*) and low (*B*) atrial septal defects in the heart. Red denotes well-oxygenated arterial blood; blue, poorly oxygenated venous blood; purple, a mixture of arterial and venous blood. Permission pending.

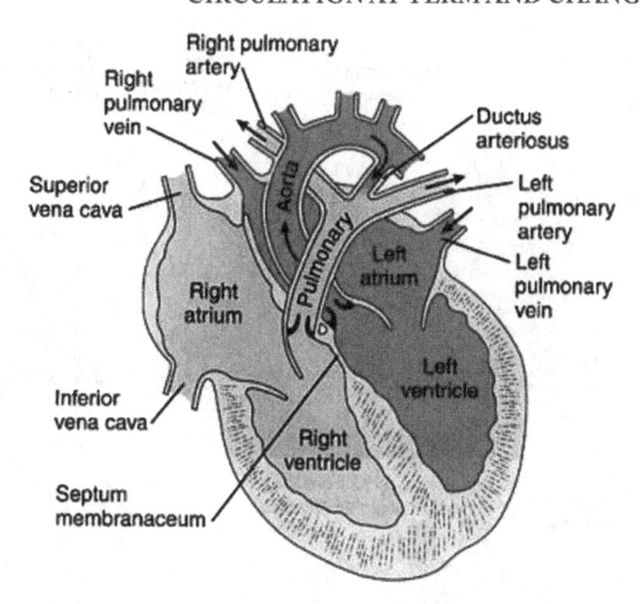


FIGURE 38: Patent ductus arteriosus showing the flow of blood from the aorta into the pulmonary circulation. Later in life, pulmonary hypertension may result, causing the reversal of blood flow through the shunt and cyanosis. Permission pending.

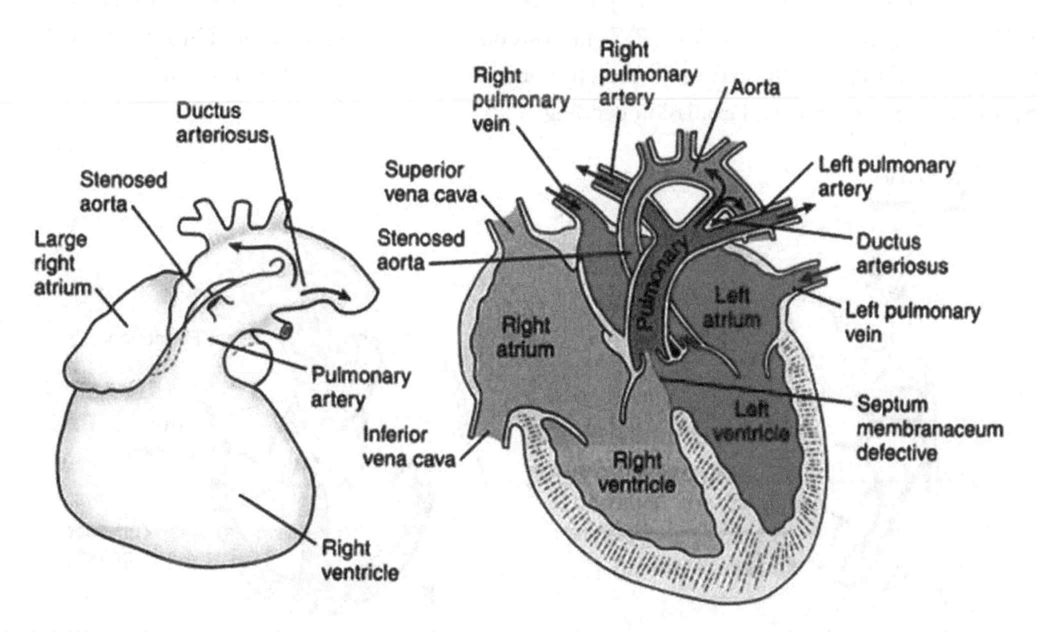


FIGURE 39: Aortic stenosis with patent ductus arteriosus. In severe cases the ductus arteriosus commonly remains patent. (*Right*). Mixed arterial and venous blood in the pulmonary artery is shown in purple. Initially, blood from the pulmonary trunk (*purple*) goes through the ductus arteriosus into the aorta, often leading to cyanosis. Permission pending.

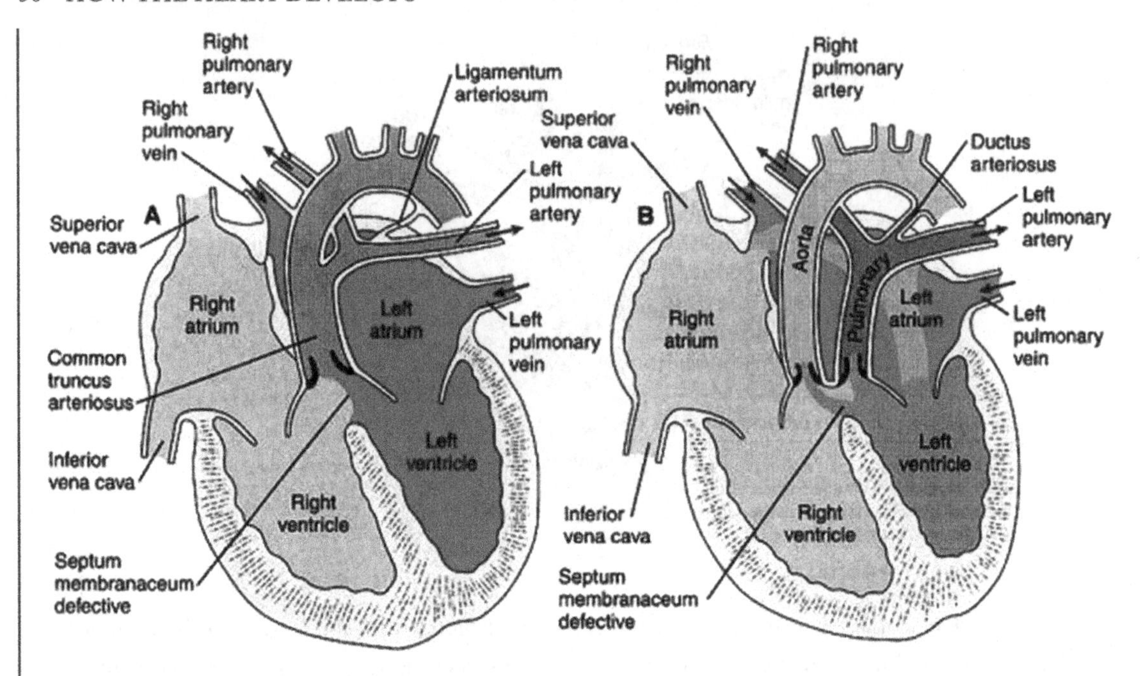


FIGURE 40: Transposition of great vessels with single outflow tract. (*A*) Persistent truncus arteriosus. A single outflow tract is fed by blood entering from the right and left ventricles. The membranous part of the IV septum is commonly defective. (*B*) Transposition of the great vessels caused by lack of spiraling of the truncoconal ridges in the early embryo. The aorta arises from the right ventricle and the pulmonary artery from the left ventricle. Permission pending.

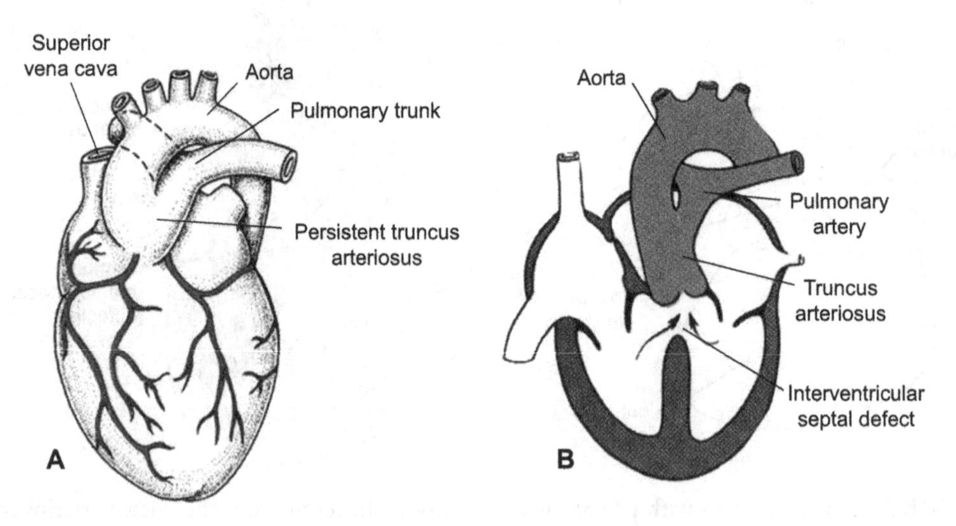


FIGURE 41: Persistent truncus arteriosus. (*A*) The pulmonary artery originates from a common truncus. (*B*) The septum in the truncus and conus has failed to form. This abnormality is always accompanied by an IV septal defect. Reproduced from Sadler. Permission pending.

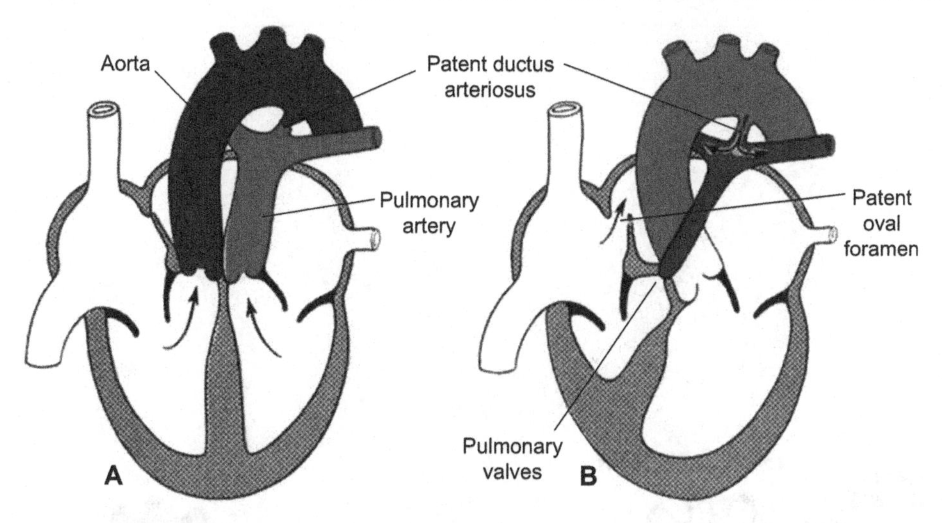


**FIGURE 42:** Transposition of great vessels and pulmonary valve atresia. (*A*) Transposition of the great vessels. (*B*) Pulmonary valvular atresia with a normal aortic root. The only access route to the lungs is by way of a patent ductus arteriosus. Reproduced from Sadler. Permission pending.

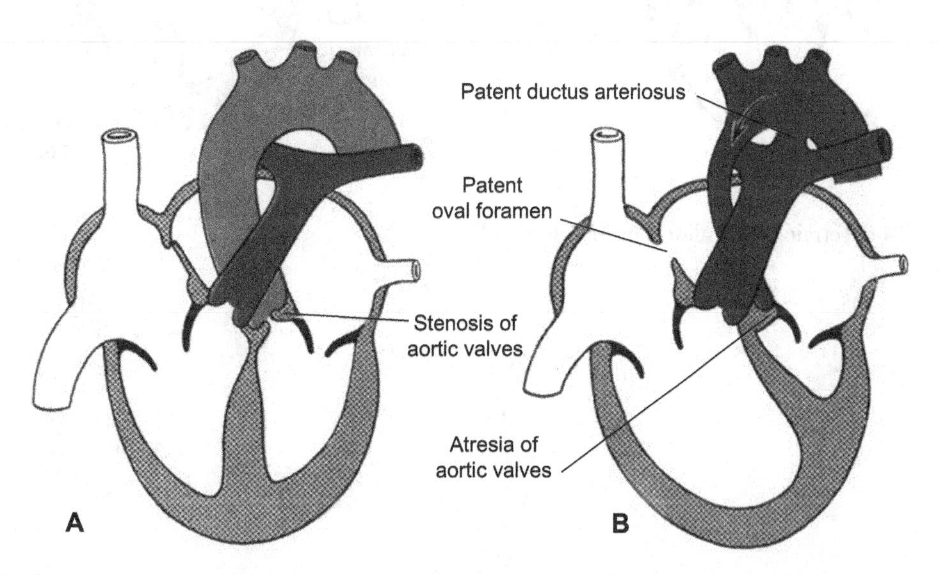


**FIGURE 43:** Aortic valve stenosis and aortic valve atresia. (*A*) Aortic valvular stenosis. (*B*) Aortic valvular atresia. (*Arrow*) in the arch of the aorta indicates direction of blood flow. The coronary arteries are supplied by this retroflux. Note the small left ventricle and the large right ventricle. Reproduced from Sadler. Permission pending.

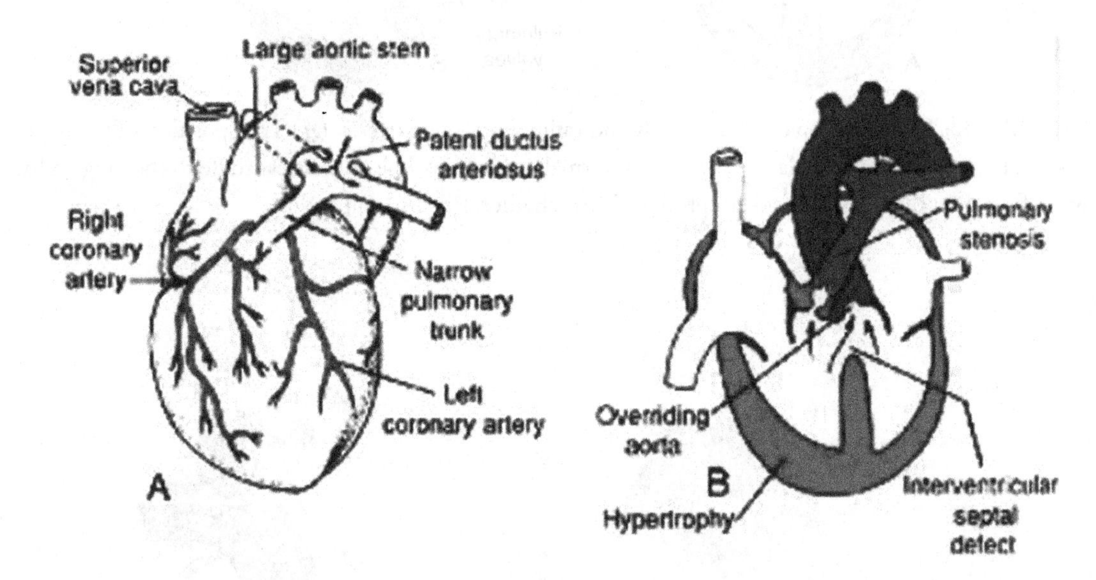


FIGURE 44: Tetralogy of Fallot. Permission pending.

# Recommended Readings

*Larsen's Human Embryology*, 4th ed. Oxford, UK: Churchill Livingstone; 2008, chapters 12 and 13.

*Human Embryology*: chapter 7 animations, http://www.med.uc.edu/embryology/chapter7/animations/contents.htm.

# Index

Amnion, 14, 20
Amniotic cavity, 20
Anterior cardinal vein, 30
Aorta, 36, 47
Aortic arch, 36, 47
Aortic sac, 18, 20
Aortic stenosis, 49
Aortic valve stenosis and atresia, 51
Aorticopulmonary septum, 35–37, 39
Aorticopulmonary trunk, 22
Ascending aorta, 37
Atrial septal defects, 48
Atrioventricular canals, 24–25, 33, 36
Atrioventricular cushions, 32
Atrium
  conducting cells in, 38
  left, 3, 18, 45
  primordial, 26
  right, 18, 46
  septation of, 26–31

Blood
  hematopoiesis, 5–8
  oxygen saturation of, 3, 45, 47
Blood islands, 5–6
Blood vessels, 6
Body folds, 11
Bradykinin, 44
Bulbar ridge, 32, 35
Bulboventricular loop, 23
Bulbus cordis, 9, 17, 30, 35–36

Carbon dioxide, 1
Cardiac jelly, 13–14
Cardiac looping, 17–21
Cardiac myocytes, 38, 40
Cardiogenic mesoderm, 13, 16, 40
Caudal body fold, 11
Cell lineages, 39–42
Chorion, 30
Circulation
  fetal, 1–3, 43, 45
  postnatal, 43, 47
Coelomic cavities, 22
Common cardinal vein, 30, 33
Conducting cells, 38
Conducting system, 38
Congenital heart defects, 48–52
Conotruncus, 18, 24
Conus arteriosus, 35
Conus segment, 17
Coronary arteries
  cellular precursors of, 13
  vasculogenesis of, 7

Coronary sinus, 46
Cranial body fold, 11
Crista terminalis, 33, 46
Cushions
  atrioventricular, 32
  endocardial, 24–25, 27, 33

Descending aorta, 47
Dorsal aorta, 30
Dorsal endocardial cushions, 24, 27, 30
Dorsal intersegmental arteries, 30
Dorsal mesocardium, 22
Ductus arteriosus, 3, 43–45
Ductus venosus, 3, 43, 45

Embryonic hematopoiesis, 5–8
Embryonic hemoglobin, 5
Endocardial cushions, 24–25, 27, 33
Endocardial tubes
  description of, 9, 12
  fusion of, 13–14, 22
Endocardium, 13–14, 23
Endothelin, 38
Epiblast cells, 9–10
Epicardium, 13–14, 23, 39

Fate mapping, 9–12
Fetal circulation, 1–3, 43, 45
Fetus
  hematopoiesis in, 5–8
  three-chambered heart of, 1–2
  Foramen ovale, 3, 26, 29–30, 43, 45–47
Foramen primum, 26–28, 33
Foramen secundum, 26–28, 33

Gastrulation, of precardiac mesoderm, 10

Heart chambers, 12, 18. *See also* Atrium; Ventricles
Hematopoiesis, 5–8
Hemoglobin, embryonic, 5
Hensen's node, 10

Inferior atrioventricular cushion, 25
Inferior vena cava, 3, 45–47
Internal iliac artery, 3, 45, 47
Interventricular septum, 27, 30, 32–33, 36

Lateral body fold, 11
Left atrium, 3, 18, 45
Left hepatic vein, 3, 45
Left ventricle, 18
Ligamentum arteriosum, 47
Ligamentum teres, 47
Liver, 5
Lungs, 43

Medial umbilical ligament, 47
Mesocardium, 22
Mesoderm
  cardiogenic, 13, 16, 40
  extraembryonic, blood cell formation in, 5
  precardiac, 9–12
  splanchnic, 13, 16
Mitral valve, 33
Myocardial cells, 25
Myocardium, 13–14, 23, 25, 42
Myocytes, 38, 40

Neural crest, 39–40
Neural plate, 6
Notochord, 14

Outflow septum, 39, 42
Outflow tract, 25, 36–37

Papillary muscles, 33
Parietal pericardium, 14, 22
Partitioning of the truncus, 35–37
Patent ductus arteriosus, 49
Pericardial cavity, 22–23
Pericardial coelom, 14
Persistent truncus arteriosus, 50
Placenta, 1
Portal vein, 3, 45, 47
Posterior cardinal vein, 30
Postnatal circulation, 43, 47
Precardiac mesoderm, 9–12, 18
Presumptive atria, 17
Primitive streak, 10, 14
Primordial ventricle, 25
Proepicardium, 13, 15, 40–41
Prostacyclin $I_2$, 44
Prostaglandin $E_2$, 44
Pulmonary artery, 37
Pulmonary atresia, 51
Pulmonary trunk, 3, 36–37, 45, 47
Pulmonary vascular resistance, 43
Pulmonary veins, 3, 45, 47

Red blood cells, 5
Right atrium, 18, 46
Right hepatic vein, 47
Right ventricle, 3, 18, 45

Semilunar valves, 35
Septation
  atrial, 26–31
  ventricular, 32–34
Septum primum, 26, 28, 33, 46
Septum secundum, 26, 29, 46
Sinistroventroconal cushion, 25
Sinoatrial orifice, 30
Sinoatrial segment, 17
Sinoatrial valve, 25
Sinus venosus, 17, 20, 22–23, 30, 33
Somite, 11, 23
Splanchnic mesoderm, 13, 16
Stem cells, 5
Superior atrioventricular cushion, 25
Superior vena cava, 3, 33, 43, 45–46
Superior vesical artery, 47

Tetralogy of Fallot, 52
Three-chambered heart, 1–2
Transposition of the great vessels, 50–51
Transverse sinus of the pericardium, 22–23
Tricuspid valve, 33
Truncal ridge, 35
Truncus arteriosus
  anatomy of, 25, 30, 35–36
  persistent, 50
Tubular heart
  components of, 17–18
  formation of, 11, 16
  looping of, 17–21
  in mouse, 16
  precardiac mesoderm transformation into, 9
  ventricular segment of, 17, 20

Umbilical artery, 30
Umbilical vein, 3, 30, 45
Umbilicus, 3, 45, 47
Urinary bladder, 3, 45

Vasculogenesis, 5–7
Ventral endocardial cushions, 24
Ventricles
left, 18
right, 3, 18, 45
septation of, 32–34
Ventricular segment of tubular heart, 17, 20
Visceral pericardium, 22
Vitelline vein, 30

Yolk sac, 5–6, 14, 20

Zebra fish heart circulation, 1, 4

# Colloquium Series on the Cell Biology of Medicine

Editor
**Joel D. Pardee**, Weill Cornell Medical College

**Forthcoming Books** (continued from page iii)

Cancer Genetics
A.M.C. Brown
2009

Cancer Invasion and Metastasis
A.M.C. Brown
2009

Cartilage: Keeping Joints Functioning
Michelle Fuortes
2009

The Cell Cycle and Cell Division
Joel D. Pardee and A.M.C. Brown
2009

Cell Metabolic Enhancement Therapy for Mental Disorders
Joel D. Pardee
2009

Cell Motility in Cancer and Infection
Joel D. Pardee
2009

**Cell Structure and Function: How Cells Make a Living**
Joel D. Pardee
2009

**Cell Transformation and Proliferation in Cancer**
A.M.C. Brown
2009

**Cholesterol and Complex Lipids in Medicine: Membrane Building Blocks**
Suresh Tate
2009

**Creating Proteins from Genes**
Phil Leopold
2009

**Development of the Cardiovascular System**
D.A. Fischman
2009

**Digestion and the Gut: Exclusion and Absorption**
Joel D. Pardee
2009

**The Extracellular Matrix: Biological Glue**
Joel D. Pardee and A.M.C. Brown
2009

**Fertilization, Cleavage, and Implantation**
D.A. Fischman
2009

**Gastrulation, Somite Formation, and the Vertebrate Body Plan**
D.A. Fischman
2009

**Generating Energy by Oxidative Pathways: Mitochondrial Power**
Suresh Tate
2009

**Heart Structure and Function: Why Hearts Fail**
D.A. Fischman
2009

**Hematopoiesis and Leukemia**
Michelle Fuortes
2009

**How Amino Acids Create Hemoglobin, Neurotransmitters, DNA, and RNA**
Suresh Tate
2009

**The Human Genome and Personalized Medicine**
Phil Leopold
2009

**Mechanisms of Cell & Tissue Aging: Why We Get Old**
Joel D. Pardee
2009

**Mendelian Genetics in Medicine**
Phil Leopold
2009

**Metabolism of Carbohydrates: Glucose Homeostasis in Fasting and Diabetes**
Suresh Tate
2009

**Metabolism of Fats: Energy in Lipid Form**
Suresh Tate
2009

**Metabolism of Protein: Fates of Amino Acids**
Suresh Tate
2009

**Neurulation, Formation of the Central and Peripheral Nervous Systems**
D.A. Fischman
2009

**Non-Mendelian Genetics in Medicine**
Phil Leopold
2009

**Skeletal Muscle, Muscular Dystrophies, and Myastheneis Gravis**
D.A. Fischman
2009

**Skin: Keeping the Outside World Out**
Joel D. Pardee
2009

**Stem Cell Biology**
A.M.C. Brown
2009

**Therapies for Genetic Diseases**
Phil Leopold
2009

**Tissue Regeneration: Renewing the Body**
Joel D. Pardee
2009

# Colloquium Series on Integrated Systems Physiology



**Forthcoming Books**

Capillary Fluid Exchange
Ronald Korthuis
2009

Endothelin and Cardiovascular Regulation
David Webb
2009

Hemorheology and Hemodynamics
Giles Cokelet
2009

Homeostasis and the Vascular Wall
Rolando Rumbaut
2009

Inflammation and Circulation
D. Neil Granger
2009

**Integrated Cardiovascular Responses to Exercise**
Doug Seals
2009

**Liver Circulation**
Wayne Lautt
2009

**Lymphatics**
David Zawieja
2009

**Ocular Circulation**
Jeffrey Kiel
2009

**Pulmonary Circulation**
Mary Townsley
2009

**Regulation of Arterial Pressure**
Joey Granger
2009

**Regulation of Cardiac Contractility**
John Solaro
2009

**Regulation of Endothelial Barrier Function**
Harris Granger
2009

**Regulation of Tissue Oxygenation**
Roland Pittman
2009

**Regulation of Vascular Smooth Muscle Function**
Raouf Khalil
2009

**Skeletal Muscle Circulation**
Ronald Korthuis
2009

**Vascular Biology of the Placenta**
Yuping Wang
2009

# Colloquium Series on Developmental Biology



**Forthcoming Books**

**Fibroblast Growth Factor Signaling**
Elizabeth Pownall and Harvey Isaacs
2009

**Formation and Differentiation of Placodes**
Jean-Pierre Saint Jeannet
2009

**Formation of the Embryonic Mesoderm**
Daniel Kessler
2009

**Maternal Control of Embryogenesis**
Florence Marlow
2009

**Neural Crest Lineage**
Patricia Labosky
2009

**Organization of the Nervous System**
Dale Frank
2009